*One Hundred Years
of Water Wars in New Mexico
1912-2012*

# *One Hundred Years of Water Wars in New Mexico*
## *1912–2012*

A New Mexico Centennial History Series Book

Edited by
## Catherine T. Ortega Klett
New Mexico Water Resources Research Institute

SUNSTONE PRESS
SANTA FE

© 2012 by New Mexico Resources Research Institute
All Rights Reserved.

No part of this book may be reproduced in any form or by any electronic or mechanical means including information storage and retrieval systems without permission in writing from the publisher, except by a reviewer who may quote brief passages in a review.

Sunstone books may be purchased for educational, business, or sales promotional use. For information please write: Special Markets Department, Sunstone Press, P.O. Box 2321, Santa Fe, New Mexico 87504–2321.

Book and Cover design › Vicki Ahl
Body typeface › Bell MT
Printed on acid-free paper
∞

Library of Congress Cataloging-in-Publication Data

One hundred years of water wars in New Mexico, 1912-2012 / edited by Catherine T. Ortega Klett, New Mexico Water Resources Research Institute.
   pages cm -- (A New Mexico centennial history series book)
   ISBN 978-0-86534-902-5 (softcover : alk. paper)
   1. Water--Law and legislation--New Mexico. 2. Water rights--New Mexico--History.
3. Pueblo Indians--Legal status, laws, etc. 4. Pueblo Indians--Claims--History.
I. Ortega Klett, Catherine T., 1956-
   KFN3723.W2O54 2012
   363.6'109789--dc23
                    2012029635

**WWW.SUNSTONEPRESS.COM**
SUNSTONE PRESS / POST OFFICE BOX 2321 / SANTA FE, NM 87504-2321 /USA
(505) 988-4418 / ORDERS ONLY (800) 243-5644 / FAX (505) 988-1025

# Dedication

We dedicate this book to the memory of Dr. Bobby J. Creel, our colleague at the Water Resources Research Institute from 1986 until his untimely death in February 2010. During his tenure, he received over 60 grants and was the principal investigator of numerous projects involving water resources management and planning in New Mexico. He was also extremely generous in applying his expertise to the resolution of thorny water related issues over the years, and in this sense he was also highly respected as an indispensable mediator. In other words, he epitomized the approach needed time and time again to arrive at settlements and agreements in the various water wars that are the subject of this book. Dr. Creel was to have co-authored our final chapter on the subject of future water wars.

# Contents

**Preface** / 9

**Acknowledgments** / 11

**1**
Water Wars During Our Territorial Years by John W. Hernandez / **19**

**2**
Adjudications: Managing Water Wars in New Mexico by Jerald A. Valentine / **29**

**3**
Ready to Fight: Steve Reynolds—Institution, Engineer, Litigator
by John W. Hernandez / **52**

**4**
*United States v. New Mexico* by Richard Simms / **66**

**5**
"A Dog of a Lawsuit": *Texas v. New Mexico* by Em Hall / **84**

**6**
*Albuquerque v. Reynolds*: Conjunctive Use and Municipal Water Supply
by Jay F. Stein / **97**

**7**
Las Vegas, New Mexico: The Rise and Fall of the Pueblo Water Rights Doctrine
by James C. Brockmann and Eluid L. Martinez / **109**

**8**
The Middle Rio Grande Minnow Wars by Charles T. DuMars / **123**

**9**
Struggle Over Pueblo Water Rights: The *Aamodt* Case by John W. Utton, and
Aamodt, Schmaamodt: Who Really Gets the Water? by John Nichols / **141**

**10**
The Writing and Filming of *The Milagro Beanfield War* by John Nichols / **180**

**11**
For the Sake of Peace in the Valley: The Negotiated Settlement in the Taos Water
Rights Adjudication by Sylvia Rodríguez / **198**

**12**
Water Transfers and the Weight of Public Welfare Considerations
by Calvin Chavez / **206**

**13**
Conflicts in the Division of New Mexico's Share of the Colorado River
by John W. Hernandez / **213**

**14**
Whose Water Is It, Anyway? Anatomy of the Water War Between El Paso, Texas
and New Mexico by Linda G. Harris / **227**

**15**
The Mother of All Water Rights Adjudications? by Kay Matthews / **254**

**16**
Future Water Wars in New Mexico by M. Karl Wood / **262**

**Epilogue** by Michael L. Connor / **283**

**Contributors** / **286**

# Preface

### John W. Hernandez

This book is about water wars in New Mexico—particularly those water-related conflicts that have arisen in the hundred years since statehood. The history and legacy of such disputes has become part of what we celebrate in our centennial. For various reasons, conflicts over water have always been with us in New Mexico. Is the crux of the problem location, location, location? We are situated on a section of the earth that has a great climate, but it also does not get a whole lot of precipitation. New Mexico is an arid state with average rainfalls that may be enough to raise a crop above 7,500 feet, and far too little rainfall to grow much, except for a few places, below that elevation. Episodic periods of drought at all elevations causes demand to outstrip supply for years at a time. Efforts to create administrative rules to manage our limited supply have been successful sometimes, not so successful other times, and frequently controversial. That is why we scrap over what little water we have.

Water fights are an integral part of our lives, maybe more so for us than for other arid-lands people. In this book, we look at some pre-1912 skirmishes to provide insight as to why New Mexico has experienced so many water-use quarrels in the brief span of the one hundred years since statehood. Many of the disputes described in the book helped create some of the tools by which water is managed in most of the western states.

We will touch only on a handful of the many large and small disagreements over the right to use water that can be characterized as "water wars." New Mexico has suffered fights with neighboring states, disagreements between Native American people and the Hispanic settlers, encounters between the state and the federal government, quarrels between upstream and downstream users, and mêlées among neighbors on the same watershed. We offer you at least one story related to each of these characteristic types of water wars.

One of the settings for water fights is in the adjudication process. Adjudications start in the courts and are like 'quiet title property suits,' except there may be thousands of water-rights holders involved, each trying to get a fair deal, and maybe more. In other cases, a water war occurs as a result of a confrontation over water rights that is of such magnitude that the courts are asked to intercede. This happens when a clash has occurred with such serious associated consequences that a resolution is unlikely to come about through negotiations, and resort to the courts is the only solution. Once a battle over water reaches the courts, there will likely be winners and losers. No one likes to be a loser. We will describe a few cases where friendly discussions, or in some instances, hostile dialogue, have produced an acceptable outcome where no one lost very much.

Water wars are typically very messy and complicated—full of legal precedents, revolving around interpretations of law that statute-framers never suspected or anticipated, and so complex that no simple narrative can provide all of the facts and ramifications. We have not asked authors to give both sides of a case—we hope that they will provide a somewhat balanced view of what happened, but we do not expect them to know or to give all the details that will please everyone involved.

In the chapters that follow, we offer fifteen short accounts of multifaceted, typically legal situations that should be of interest to all New Mexicans who love a story about a good fight. How did the disagreement get started? Why? When? Who were the players? What were the outcomes? How will, or did, these results change over time? And who were the winners and who were the losers? Without too much legalese, we hope that we have answered these questions in each of our fifteen accounts.

**Causes of Water Wars Before and After Statehood**

The concepts behind Chapters 1, 2, and 3 are intended to answer the question: Why have we, the people of New Mexico, had so many conflicts over water? Chapter 1 discusses the impact of our long colonial and territorial periods, and our tendency toward quarreling—with everyone in sight—over our rights to use water. This chapter postulates that we are more inclined to fight over water than folks in other western states because of the way this area was settled, and the many years that it took for New Mexico to become a state. Management of our water resources during our years as a territory didn't help, and may have set the stage for some of today's water wars.

In Chapter 2, Judge Jerald A. Valentine gives his perspective on how the courts have interpreted the Water Code and Rules of Civil Procedure trying to expedite resolution of contentious legal and factual disputes while maintaining due process for water right claimants. Judge Valentine notes that "contentious water disputes have been with us for a long time and will continue in the future." Most of our authors also reflect that judgment.

Chapter 3 is about longtime State Engineer, Steve Reynolds, who guided our water ship through troubled seas for 34 years. Many of the authors of our stories worked for him, as I did. Was State Engineer Reynolds overly aggressive in protecting existing water rights or was he just doing what the laws required? I think Reynolds may just have liked a good fight. Why else did he keep track of how many court cases he won and lost over his long-tenured position of power?

**Water Cases that Reached a Supreme Court**

Our second set of centennial stories includes four cases that reached either the New Mexico Supreme Court or the U.S. Supreme Court. State Engineer Steve Reynolds, a man you will hear a lot about in this book, always used to say, "The Supreme Court may not be right, but it is always supreme!"

The first of our U.S. Supreme Court accounts is by Richard Simms, a western water rights expert and a lawyer for Reynolds' Office of the State Engineer (OSE) in the 1970s and early 80s. Mr. Simms appeared before the U.S. Supreme Court in *United States v. New Mexico*, a landmark case that firmly established the limits of federal reserved rights on the public domain in western states. He will tell you about the threat posed to New Mexico had the federal government prevailed in this case. Chapter 5 is a U.S. Supreme Court story about New Mexico's efforts to deliver water to Texas via the Pecos River and the lawsuit involved. Professor Emeritus Em Hall, at the University of New Mexico's School of Law, will tell you about the problems associated with the Pecos River Compact, and how it led to our state line water delivery problems. Em is the author of a book on the subject entitled, *High and Dry: The Texas-New Mexico Struggle for the Pecos River*.

The next two chapters concern New Mexico Supreme Court cases. In 1956, when Steve Reynolds declared much of the Middle Rio Grande watershed to be a groundwater basin subject to control by his office, there were many who thought that Reynolds would soon be returning to his previous job as a researcher at New Mexico Tech. The City of Albuquerque, particularly reliant on groundwater for its future growth, sued and sought relief from Reynolds' declaration in the state's highest court. Among other issues, Albuquerque claimed that its future use of the area's groundwater was protected under the Pueblo Water Rights Doctrine. Jay Stein, an ex-Office of the State Engineer and Interstate Stream Commission lawyer and now a Santa Fe water lawyer, will tell you what happened and how much winning meant to the power of State Engineer Reynolds.

Chapter 7 is by James C. Brockmann, a water attorney in private practice, and by ex-State Engineer, Eluid Martinez. As a young engineer, Martinez worked for Reynolds for years and was the hearing officer on many water rights cases. Some New Mexicans believe that the Treaty of Guadalupe Hidalgo provided support for the Pueblo Water Rights Doctrine. Under this concept, a community that had been determined to be a pueblo prior to the Mexican-American war could take as much water from a river as was necessary for its present and future inhabitants, now and forever into the future. The New Mexico Supreme Court made this finding for the Town of Las Vegas in 1958. State Engineer Martinez challenged the right of the Town of Las Vegas to have preferred access to the area's water resources forever. In their account, Brockmann and Martinez will tell you what happened when the New Mexico Supreme Court re-examined the Pueblo Water Rights Doctrine in our state.

## A Case that May Go to a Higher Court Someday

Certainly not all water quarrels make it to the highest courts in the land, but there are some serious issues that may make it there some day. Chapter 8 is about water problems associated with the federal application of the Endangered Species Act in New Mexico. Charles T. DuMars, Professor Emeritus at the University of New Mexico, now in private practice, will tell you of the rounds of fighting that have taken place in U.S. District Court between the irrigators of the Middle Rio Grande valley, who hold water rights, and those who believe the Endangered Species Act demands that the Rio Grande silvery minnow be given all the water needed for its survival.

## Four Stories Involving Acequias

Depending on who is counting, there are still almost 800, small, locally managed, community irrigation systems in the state. At the time of statehood, the acequias represented the bulk of the irrigated farmlands in New Mexico.

Acequias have their own state water laws because these existing community ditch systems just didn't seem to fit under our 1907 Water Code. Their management is unique and the legislature has tried to maintain the acequias' historic customs. Changes made in state statutes in 2003 make it very difficult to transfer water rights from a community ditch to some purpose other than irrigation.

Chapter 9 is about the Aamodt case that holds the distinction of being the longest pending case in the federal court system. The case was filed in 1966 by Steve Reynolds as a water rights adjudication of the Pojoaque Basin north of Santa Fe, including the Rio Pojoaque and its two major tributaries, the Rio Tesuque and Rio Nambe. Edwin Mechem, an ex-Governor, ex-U.S. Senator, and longtime federal judge presided over the Aamodt case for years, and many hoped that he would be the last judge on the case. When he died in 2002, he left the case near settlement, ten years later, Aamodt is finally near solution.

We give you two accounts of the Aamodt case. The first is by John W. Utton, a lawyer representing Santa Fe County interests, who has been a participant in the case for many years. John tells the story of the water use conflict that started well before statehood between the Pueblo Indians and the Hispanics and Anglos who settled later, and who obtained water rights for their community ditches by filing applications with the New Mexico State Engineer for the appropriation of water. John Utton's comprehensive account takes the reader from the territory years of land-use differences between the Pueblo people and the settlers to the present time and the 2010 Congressional Act that has made a final settlement of the case possible.

Our second account of the Aamodt case is a contemporary view of the lack of community support for the proposed settlement. It is written by John Nichols, author of a number of stories about life on northern New Mexico's community ditches. His most famous story is *The Milagro Beanfield War*. In his Aamodt account, Nichols tells the reader some of the reasons why the congressional settlement has not been viewed with joy by all.

Chapter 10 is also by John Nichols. He recounts the delays, uncertainties, trials and tribulations associated with Robert Redford's filming of the Beanfield War book. He also gives the reader insight into his many picturesque characters in his story. Nichols' book is about the use of an old acequia to water a local bean field. The conflict between the local farmers and a recreational development company started in 1955 when the U.S. Bureau of Reclamation proposed to use part of the San Juan-Chama water allocation to provide a more reliable water supply for the farmers on community ditches below Ranchos de Taos, but the Reclamation project would also create two new dams and add new water that could be used for recreational purposes.

Chapter 11 is another Taos story of the conflicts between the Pueblo people and the local settlers and the claims that each makes to the use of the area's streams. Dr. Sylvia Rodriguez, author of the book *Acequia*, and professor emerita from the University of New Mexico, tells you about the negotiated settlement between Taos Pueblo and the area acequias that has been made possible by the Congressional Settlement Act of 2010. A settlement will bring final consensus in the adjudication of Taos area water rights that have been on hold for almost 40 years.

Chapter 12 is an unusual acequia case that involved the transfer of water from the northern New Mexico Ensenada Community Ditch to a proposed recreational development. State Engineer Steve Reynolds originally approved the transfer. The transfer was protested by the people in the local community and the case went to trial before District Court Judge Art Encinias in Tierra Amarilla. Calvin Chavez, a veteran engineer with the Office of the State Engineer, will tell you about what happened in that courtroom and subsequently in the New Mexico Court of Appeals. Unfortunately, Mr. Chavez died before finalizing his story and a foreword and afterword are provided to fill in some of the gaps.

**Battles Over Water in Our Larger Rivers**

The San Juan River in northwestern New Mexico is one of the state's shortest, but it carries more water than any other. The source of the water is snowmelt from mountains on the western slopes of Colorado. The San Juan is our major tributary of the Colorado River system. Chapter 13 is about two federal projects designed to utilize New Mexico's share of the Colorado River water: the Navajo Irrigation Project and the San Juan-Chama Diversion. Both projects involve water from the tributaries of the Upper Colorado River and both projects have been built and are operated by the Bureau of Reclamation. In the early stages, when Ed Mechem was governor in 1953, he linked the two projects together, stating that he would support both if a final agreement on the water rights of the Navajo Nation was reached. After 50 years of development, the Navajo Irrigation Project is yet to be completed and final settlement of Navajo water claims is still to come. Fortunately, the San Juan-Chama Project has been completed and Albuquerque, the principal recipient of San Juan water, is now taking the major part of its public water supply from the diversion project. We will tell you about issues related to these two projects and where the situation now stands.

In Chapter 14, Linda G. Harris writes about the El Paso case. Harris, author, historian, and former editor at the New Mexico Water Resources Research Institute, covered the 54-day El Paso hearing in 1986 and 1987. The case began in 1980, when the City of El Paso claimed the right to appropriate a large amount of groundwater from an undeclared basin in New Mexico. Steve Reynolds, who was the hearing officer, ruled that El Paso did not get the right to pump groundwater from the Mesilla Bolson. Appeals and challenges to New Mexico water law were filed following the ruling. Aspects of the case lingered until a settlement between the two area irrigation districts, one in Texas and one in New Mexico, was reached in 2008. The settlement determined how Texas would receive its share of the water from Elephant Butte Reservoir.

In Chapter 15, Kay Matthews, a journalist from the northern part of the state, tells you about "the mother of all water wars." It is a classic story of intrigue that started in pre-statehood days and is now in Round 2 where battles will continue for years to come. Round 1 of this Rio Grande water war is about conflicts that involved Mexico, Colorado, Texas, New Mexico, the U.S. Supreme Court, the U.S. Congress, and the U.S. Secretary of State, Secretary of Interior, and the U.S. Attorney General. It is a compelling story about Dr. Nathan Boyd's 1894 plan to dam the Rio Grande at Elephant Butte and to build diversion dams and canals to irrigate lands in New Mexico, Texas, and maybe Mexico. The Boyd plan was finally put to rest in the U.S. Supreme Court in the middle 1920s. Was Round 1 a great deal to do about something that ended in nothing—or not? Round 2 is about what happens next on the Boyd plan as Kay Matthews will tell you.

Our final Chapter 16, by Dr. Karl Wood, former director of the New Mexico Water Resources Research Institute, provides perspective on the water wars New Mexico will face in the future. The recurring theme of this chapter is that with respect to water wars, the past is prologue. Many of the major and minor water disputes discussed earlier in the book are reviewed again, and it is argued that in general for water wars there are few, if any, final victories, but rather a succession of uneasy temporary settlements, agreements, and truces. It is also made clear this happens in part due to the great complexities and ambiguities surrounding these historical disputes and in part because water is an increasingly scarce resource. Of course, the underlying fear is there may not be enough available at low enough cost to satisfy the growing needs of all the competing interests. And so this chapter makes the persuasive case that future water wars will naturally evolve and erupt out of the exiting status quo for water in New Mexico.

Current Bureau of Reclamation Commissioner Michael L. Connor, graciously provides an epilogue to our book. He hopes that "as our water management challenges get tougher in the future, New Mexicans will pull together and make decisions that will allow us to continue to thrive and prosper in this great state." That will indeed be the key determinant of success in this debate going forward, and it will require great constancy of purpose to that end. Only time will tell over the next hundred years to what extent the desired results are achieved.

# Acknowledgements

Dr. James D. Klett, Will Keener, and Jennifer E. Fletcher spent many hours reading several versions of each chapter and their comments and attention to detail is very much appreciated. This book would not have come to fruition without the participation of John W. Hernandez, Professor Emeritus, New Mexico State University. Dr. Hernandez came to the New Mexico Water Resources Research Institute with the idea for the book, prepared a book proposal for consideration by the Centennial History Series, contacted most of the authors asking for their participation, and wrote several chapters. We thank him for his encouragement and willingness to push forward on the project.

# 1

# Water Wars During Our Territorial Years

John W. Hernandez

**Is New Mexico More Subject to Water Wars than Other Western States?**

There are some of us who believe, without the necessity of a resort to scientific study or to rational thinking, that as westerners we are more prone to be involved in water quarrels than folks who live east of the Rockies. Everyone in the West knows that "Whiskey's for drinking; water's for fightin'." Given that that's true, are water users in New Mexico more inclined to water wars than the rest of the West? I think so, for a flock of reasons. I'll tell you about some of them in the sections that follow.

**Is Our Hispanic Ancestry a Root Source of Our Water Wars?**

While not the fundamental problem, the answer is probably, yes. Our heritage of Iberian customs, in the management of scarce water resources, is certainly a contributing factor in our apparent tendency toward water conflicts. One look at the snow capped mountains and arid, fruitless plains of Andalucía is enough to convince a New Mexican that, yes, this is where many of us came from, followed by an unvoiced certainty that water practices that were used in Spain came with us to New Mexico.

It should be noted that some parts of Spain followed slightly different water codes than others. Some regional differences prevailed. As southern Spain was the last stronghold of the Moors, we probably also inherited some of their customs and technology in designing and managing the early community ditches in New Mexico, the acequias, that were the backbone of much of the farming in our territorial days.

Elements of Spanish and Moorish practices that made it into New World water codes included: the ownership of water in a river belonged to the general public for their free use; the rights of existing water users to divert water from a stream were protected; the rights to use water were tied to the land where application was made; canal systems belonged to those who built them, and right-of-way to ditch for construction and maintenance was guaranteed; these ditch owners annually prescribed their own rules for scheduling cleaning and maintenance of the ditch and the times, amounts, and methods of water diversions from the acequia to farm fields; water use was limited to beneficial purposes; and limits were placed on upstream diversions to that which was absolutely needed. In some areas, constraints were probably imposed on developments of springs, seeps, and shallow groundwater.

## Were Our Years as a Province of Spain a Source of Future Conflicts?

The application of Spanish water use traditions in New Mexico was complicated by the fact that there were groups of farmers already here, using the scarce waters of the state, when the first settlers arrived. In the northern part of New Mexico, they were the Pueblo Indians. We now had Moorish irrigation technology and Spanish water customs overlaying Pueblo Indian farming practices. Spanish settlers also encroached on the already limited water supply used by the Pueblo Indians and settled in some Indian occupied lands. The consequences of conflicts between Hispanic and Indian farmers remain with us today. It has been necessary for the federal government to fund costly settlement suits, the latest in 2010, when money was made available to resolve the Aamodt and Abeyta water adjudications.

Prior to our territorial years, Spanish and then Mexican authorities protected Pueblo water rights by giving them first right to water use when acequias and Pueblo Indians used water from the same streams. But conflicts occurred in territorial years as the Hispanic farmers occupied more and more land along streams used by the Pueblos. The concept of "first in right to water" was not followed on some acequias as priorities were nearly impossible to determine. Some followed the practice where the mayordomo apportioned water based on equitable sharing by all and not on a "first in right" priority basis. The Pueblos were allocated the amount of water needed to grow their traditional crops.

When the early settlers came to the Spanish Province of New Mexico, they began farming using water from the smaller tributaries of the larger rivers for irrigation because they didn't have the technology nor the equipment to build the types of diversion dams we now have on major rivers. Communities were located far apart and there was no single central authority that could, or that wanted to, exercise control over all water use in the Province.

Unfortunately, when the water-use practices came with the Spanish to the Americas, no comprehensive legal codes were adopted. Instead, a mass of laws, rules, and ordinances existed that were limited to the management of local water resources. Variations in regional rules and court decisions were made to facilitate local use and to resolve differences among water users on different tributaries of the same stream.

Even though we inherited a general set of water management rules from the Spanish, a number of things complicated the administration of water resources. One was that our river systems are very long and the state is relatively large. The Rio Grande and its tributaries drain the entire central part of the state from north to south (500 miles of river), making the effects of upstream uses on older downstream diversions and uses difficult to anticipate and impossible to quantify.

Spanish settlement of the state took a very long time (over 250 years) and small isolated communities, many in mountainous areas, characterized our development process. Some of us want to believe that some of the early settlers were Jewish and that they sought farming in isolated mountain retreats as a way of escaping domination by the Catholic Church. I share that belief and probably that ancestry. I think that moving to "the ends of earth" made some of these people not only secretive, but difficult and ready to fight for their right to use water. But that characteristic was common to most of the folks in the state in those early days. Steve Reynolds, our longtime state engineer, once noted that "Historically, many water rights problems were settled with a shovel or a shotgun. Today those means haven't been abandoned but are less frequently applied." Steve was right; our cussedness prevails, at least when it comes to water.

Local water shortages occurred on many of the acequias as there were no water storage reservoirs, and the supply from snowmelt and rainfall varied greatly from year to year and as a result was quite unpredictable. Quarrels over water between neighbors and neighboring acequias were common. Conflicts between Native American farmers and Spanish settlers also flared.

## Did Our Years as a Territory Contribute to Water Conflicts?

Were there water wars during our 75 years as a territory? Yes, you bet there were. One of the territorial period water conflicts, and one that has been dubbed by some as the "mother of all water wars," is the subject of Chapter 15 in this book. Did the protracted territorial years of poor water management leave us with conditions—situations—that made us prone to fighting over the limited supply that we have? I think so, but you will have to read on and make your own finding.

New Mexico already had hundreds of acequias (those of both Spanish and Native American farmers) using the water from almost all of the streams in the Province when American settlers from the East began to arrive after 1830. For the most part, the existing farmers followed Spanish water use practices. General Stephen Watts Kearny is reported to have announced the territory's first water code in 1846. The Kearny Code said that all of the laws and customs of land ownership, and water use, would remain in effect. In a 1984 speech, State Engineer Steve Reynolds noted that the First Territorial Legislature enacted a law providing that all of the inhabitants had the right to construct either private or community acequias to take water from any source they could. Reynolds made no mention of the lack of control of appropriations by the territorial government that would lead to future conflicts as there had already been prior appropriation by downstream users of some of the water that would be later developed by upstream users under the 1851 Act.

Water management in the territorial years was complicated somewhat by the idea that somehow the territorial government was supposed to take into account the terms of the Treaty of Guadalupe Hidalgo, the accord that ended the war of 1848 between the U.S. and Mexico. The U.S. agreed to respect the property rights belonging to "present owners, heirs of these [properties], and all Mexicans who may hereafter acquire said property . . . as if said property belonged to citizens of the United States." This meant that all prior appropriations of water would be recognized. Unfortunately, the territorial government did not have the legal structure in place until 1907 to keep track of appropriations of water prior to 1848 or after.

The difference between "property ownership," as guaranteed by Guadalupe

Hidalgo and "water right ownership" was also a problem. During our Spanish colonial years, there were no distinctions made between property ownership and water-right ownership. Under the Spanish system, if you were an irrigator and you had property rights, then you had the right to use water. The Treaty of Guadalupe Hidalgo did not speak specifically about water rights, but clearly, water rights were included in the concept of land ownership in the treaty, as was the doctrine of prior appropriation.

**Water Related Acts of the Territorial Legislature**

Most of the irrigated agriculture in New Mexico in the 1850s involved community ditch systems, or acequias. Each ditch had a "water boss," the "mayordomo," probably the most important man in the community. Different acequias had different rules. The territorial legislature was kept busy trying to figure out how best to operate the 700 or 800 community ditch systems. They should have left well enough alone. The legislature specified, and then changed, the methods of electing the mayordomo of acequias. Other legislative acts gave greater authority to the mayordomo of the "acequia madre," the mother ditch that fed a number of smaller acequias than to other ditch bosses.

During the territorial years, there were water conflicts in areas where a number of acequias took water from the same river system. These quarrels were so convoluted that the legislature repeatedly intervened by providing specific directions for the operation of the warring acequias. For example, in Taos County, a territorial measure required that the Rio Chiquita be kept flowing at all times, night and day, summer and winter. In legislation related to Rio Arriba County, the number of days (and nights) for irrigation from various acequias was set as follows: the upper El Rito acequia, ten days, the En Medio and Espinosa ditch, as far as the fields of Manual Martin who was deceased, four days each, and the Los Atencios and Los Lopez acequias, two days each. These laws further specified the number of days of labor each farmer was to provide in cleaning and maintaining the acequias. The legislature assigned the local Justice of the Peace the task of overseeing the distribution of the water between acequias and enforcing maintenance rules. Lots of luck!

Quarrels between ditches were common. The need for some of the post-statehood decisions by counties became apparent before 1912. A territorial era adjudication of water rights on tributaries of the Cimarron River in Colfax

County by a county judge took place on July 14, 1917. The county court set an 1860 priority date for the development of some of the uses on the Rayado River. Unlike many other adjudications in the state that have been done and redone, Colfax County Cause No. 4482 is still recognized by the courts and the State Engineer.

Nothing but trouble happened when the territorial legislature passed a very long bill in 1887, allowing the formation of private corporations to build dams and reservoirs, pipelines, supply canals, and laterals to irrigate newly developed lands and/or to provide a municipal water supply. These corporations could divert water from a stream as long as "surplus water" was available; that is, water that was not needed by existing irrigators. The act was explicit: no diversion of "the usual and natural flow of any stream" used by an existing acequia between the 15$^{th}$ of February and the 15$^{th}$ of October each year was allowed, nor could a new water corporation interfere with the water rights of any existing user. Who was to decide what "surplus water" was? The act didn't specify, but it allowed existing water right holders to appeal to district courts if a petition was filed in a timely fashion. The most serious problem was that no statewide authority or agency was assigned the duty of maintaining the records of existing users, or the diversions of water by newly formed corporations. Problems were sure to come, and they did.

## Supreme Court Decisions in the Territorial Years

Any time that a supreme court gets involved in a water fight you know three things: the court decision in the case will take a long time in coming; no one will be fully sure what the court's opinion says; and no one will be happy with the outcome no matter what you might think the decision says. As Steve Reynolds once observed, you may disagree with Supreme Court decisions, but "they are Supreme."

Both the territorial and U.S. Supreme Court made decisions in New Mexico water cases during our pre-state period, sometimes on the same case, but with conflicting opinions. One such incident occurred when the Territorial Court (1893) declared a series of arroyos, 4 to 18 miles in length with pronounced channels, to be "water courses" in the context of territorial law. In 1897, the U.S. Supreme Court found these same arroyos to be "merely passageways for rain" and not natural water courses. A 1912 State Supreme Court decision involving a different drainage found that, although dry most of the time, arroyos carried floodwaters during times of rain and were thus natural water courses.

Another disagreement between the territorial and U.S. Supreme Court involved the lower Rio Grande, and was part of the conflict in the "mother of all water wars" described in Chapter 15. In 1898, the Territorial Court found that the Rio Grande was not navigable within the limits of the territory, and that the waters of the river were subject to appropriation. The next year (1899) in the same case, the U.S. Supreme Court ruled that while a state could change the common law of "riparian rights," without specific action by the Congress, a state cannot destroy the rights of the federal government to the continued flow of water through property of the U.S., and that the navigability of the Rio Grande had to be maintained.

In 1897, the U.S. Supreme Court got involved in a New Mexico case concerning the concept of "pueblo water rights." The City of Santa Fe claimed to enjoy pueblo rights. The U.S. Supreme Court said, "No." In 1938, the New Mexico Supreme Court revisited the U.S. Supreme Court's decision to confirm that Santa Fe did not, in fact, have a pueblo water right. Under California rules, a pueblo right was derived from a grant from a Spanish king that gave a pueblo the right to all of the water resources of an entire non-navigable stream system. The right was open-ended and limited only by the available surface and groundwater of the stream system. A pueblo right was superior to all other rights, and not limited to uses within the original community boundaries. When faced with a 1959 New Mexico Supreme Court decision that the Town of Las Vegas enjoyed pueblo rights, then State Engineer Steve Reynolds suggested that California courts fabricated the pueblo doctrine to provide water for its growing cities. In Chapter 6, Jay Stein describes how the New Mexico Supreme Court reacted to a pueblo water rights claim by the City of Albuquerque in the early 1960s. In Chapter 7, attorney James Brockmann and former State Engineer Eluid Martinez talk about Las Vegas' pueblo rights and subsequent action of the New Mexico Supreme Court to revise its earlier decision. Yes, our supreme courts have differed in the past and even changed their minds, still, they reign supreme.

Did the territorial courts really recognize the terms of the Treaty of Guadalupe Hidalgo? They may have in some cases, but when the treaty was an issue in one water rights battle during our pre-statehood years, perhaps not. In 1900, the Territorial Supreme Court found that the U.S., and the territory both had systems of laws with respect to water rights, and that "these laws must govern wherein they differ from treaty provisions, and where they are harmonious, treaty provisions need not be considered." Huh? Well, so much for the Treaty of Guadalupe Hidalgo in territorial New Mexico!

It was in a complicated case (as are all water cases) in 1899, when an irrigation company proposed to build a canal system from below San Felipe Pueblo to Albuquerque to irrigate 7,000 acres of new lands using available surplus water, that is, water in excess of prior appropriations of water. There was prior use of northern Rio Grande water by ten existing ditches that would be crossed by the new canal. Many existing downstream users existed on the Rio Grande in southern New Mexico, Texas, and Mexico. Costly appeals by area acequias were filed. The case went to the Territorial Supreme Court, and after a hodgepodge of hocus-pocus hydrology of Rio Grande streamflows, the Court found that surplus water was available, even though serious shortfalls in needed water supplies had occurred since the middle 1880s at Las Cruces, El Paso, and Juárez. The Court decided that the prior right of existing downstream irrigators need not be considered as they were not parties to the case under consideration. The downstream folks had failed to enter an appeal. This was an interesting conclusion that did not lead to goodwill between northern and southern parts of New Mexico.

Noting that conflict between water users were becoming common, the doctrine of prior appropriation began to appear in territorial law in 1891 when the legislature required that a notice of construction of water control works had to be filed and that no priority of use could be established until that notice was filed. In 1898, the Territorial Supreme Court said that the doctrine of prior appropriation "is and always has been the law of the land." This was important as the decision firmly moved New Mexico from being a "riparian rights" state—just because water ran through your property you had no right to use it—to being a state were water belonged to the public until appropriated for beneficial use.

## The Good and the Bad—Our 1907 Water Code

In 1907, the Territorial Legislature enacted the first comprehensive water code for New Mexico. This code essentially persists today, even after more than 100 years of legislative diddling. Most important was the creation of the Office of the Territorial Engineer (in 1905) to oversee the appropriation of the surface waters of the state.

These aspects of the 1907 Code are the "good" consequences: the law mirrored Hispanic practice—for an appropriation to be permitted, water had to be put to beneficial use; the rights of all existing water appropriations were to be recognized; the doctrine of prior appropriation applied, first in use, first in right; water used for irrigation had to be associated with a specific track of land; a point of diversion from a stream was required; a water-right holder could only divert the amount of water needed for intended beneficial purposes; and all unappropriated water in New Mexico's rivers and streams belonged to the public and was subject to appropriation for beneficial use.

Now, for the "bad" aspects of the 1907 law. The act was silent about groundwater, seepage-water, springs, and water that came from "an unknown source." The failures of the 1907 law to address groundwater soon lead to a disagreement where the Court found that the Territorial Engineer was not authorized to allow the appropriation of seeps and springs. Trouble was afoot, but the worst was yet to come.

While the 1907 Code is still the foundation of our present water laws, what happened immediately after its passage has been, and will continue to be, the cause of many of our water wars. I blame Territorial Engineer Vernon L. Sullivan for the run on the bank that took place. When you read the First Biennial Report of the Territorial Engineer, 1907–1908, you will find the problem. The new law opened the floodgate for requests for new appropriations of water, and the Territorial Engineer granted many. From May 17, 1907 to December 1908, the Territorial Engineer received requests for appropriation of water for over 2,000,000 acres of newly irrigated lands. Without an adequate survey of prior appropriation of water by the many acequia systems already extant, with scanty streamflow records, and without adequate administrative procedures in place to determine the effects of new appropriations, the Territorial Engineer approved the appropriation of water for 700,000 acres of these two million acres in 1907 and 1908.

The list of requests for new appropriations in the First Biennial Report covers four pages of small type. Requests came from every county in the state, from private individuals and water development companies on creeks I have never heard of—Tortilla Creek, anyone? The First Biennial Report did not include already approved requests by the Reclamation Service: 20,000 acres in Carlsbad; 19,000 acres on the Hondo; 10,000 acres at Las Vegas; 180,000 acres in the Rio Grande Project; and 60,000 acres at Urton Lake, another project of which I have never heard. Where was Urton Lake? Did the project go away? What happened to the planned water supply?

In the Second Biannual report by the Territorial Engineer, Vernon Sullivan noted that everyone requesting a new water appropriation tended to overestimate greatly the water supply available to them. I think that Engineer Sullivan was one of those overestimating the surface supply in the state when he accepted and approved applications for irrigation of 700,000 additional acres. Yes, the absence of long-time measurements of streamflow was certainly a problem; one that would lead to future water conflicts.

As statehood approached in 1912, Territorial Engineer Sullivan predicted that within the following ten years, irrigated acreage in New Mexico would grow to four million acres. If the average consumptive use of water by an acre of croplands is two acre-feet, that would mean that we would need eight million acre-feet of water per year to serve Sullivan's four million acres. That's twice as much water as Steve Reynolds calculated we diverted from all sources of water for all uses in the state in the 1980s. I think that a true accounting of the acre-feet of water involved in all these 1907–08 applications that Territorial Engineer Sullivan endorsed will show that the state's surface water supply, in a couple of basins, was already over-appropriated by the time we became a state. This has lead to many of our water wars: too many claimants, and too little water, with a too unpredictable supply. Yes, our many years as a territory has contributed to our willingness and, more important, our need to fight for our rights to water.

# 2

## Adjudications: Managing Water Wars in New Mexico

Jerald A. Valentine

**Introduction**

> Human life, as with all animal and plant life on the planet, is dependent upon water. Not only do we need water to grow our food, generate our power and run our industries, but we need it as a basic part of our daily lives—our bodies need to ingest water every day to continue functioning. Communities and individuals can exist without many things if they have to—they can be deprived of comfort, of shelter, even of food for a period, but they cannot be deprived of water and survive for more than a few days. Because of the intimate relationship between water and life, water is woven into the fabric of all cultures, religions and societies in myriad ways.
> —Len Abrams, World Health Organization

War over water became common when humans took up living in settled agrarian communities. Ancient Sumer was one of the earliest civilizations. Archaeological records show cultural continuity from 5200 BC to 4500 BC. Sumarians were among the first to develop irrigated agriculture. They developed a social organization and a technology that enabled them, through their control of water to survive and prosper in an arid land.

Sumerian agriculture depended heavily on irrigation. These ancient farmers used canals, dikes, weirs, and reservoirs to maximize the use of limited water for irrigation. The canals required frequent repair and continual removal of silt and the government required individuals to work on the canals. For nearly one thousand years, Sumerian city states continually fought wars with each other over water rights. Several millennia later, New Mexico is addressing essentially the same issues faced by the ancient Sumerians.

This chapter explains water law in New Mexico and recounts how the Third Judicial District Court has managed the Lower Rio Grande Adjudication.

**Water: Plentiful and Scarce**

About 70 percent of the planet is covered in ocean, and the average depth of the ocean is over three thousand feet. Ninety-eight percent of the water on the planet is the salt water of oceans. About two percent of the planet's water is fresh. Most living things, except for life in the seas, need fresh water.

The earth is in a continual hydrologic cycle. Water occurs in nature in three physical states: water vapor, liquid water, and ice. The hydrologic cycle of the earth may be considered a closed system where the total amount of water on a planetary scale remains essentially the same. Energy from the sun causes some of the liquid and ice water on the surface of the earth to change state to water vapor.

When climatic conditions of ambient temperature, pressure, and humidity combine in the proper proportions, water vapor-laden air (air with high humidity) changes state to liquid water or ice that falls to the earth as rain or snow. Most of the rainfall and snow melt eventually returns to the oceans and seas, where the cycle is repeated.

When precipitation returns liquid water to the land surface of the earth, or snow and ice melt at elevations above the oceans, the water takes the path of least resistance as it flows downhill. If the geologic condition of the land on which the precipitation flows is impermeable, the path of least resistance will be primarily on the surface. If the geologic condition of the land is permeable, some of the water ordinarily will result in surface runoff with some water sinking beneath the surface and continuing underground as part of the entire stream system. Water located underground is commonly called groundwater. Groundwater that flows underground as a part of a stream system is hydrologically connected to the part of the stream that is flowing on the surface.

Reduction of the surface water by diversion to farmlands or other beneficial uses may, over time, impact the groundwater, and reduction of the supporting groundwater underlying a stream by pumping wells may, over time, impact the surface water. The amount of time it takes for reduction in one to affect the other depends on the amount of water removed from the stream system, the slope of the terrain, the permeability of the soil (the resistance of the soil to flow), the distance from the location of the sites where the water is diverted, and loss of water from the stream through evapotranspiration. Evapotranspiration is the water lost to the atmosphere by two processes: evaporation and transpiration. Evaporation is the loss of moisture from open bodies, such as lakes and reservoirs, wetlands, bare soil, and snow cover. Transpiration is the loss of moisture from living plant and animal surfaces.

Because of the resistance of the soil, the flow rate of the groundwater portion of the stream will be slower than the surface portion. Both the surface water and connected groundwater are recharged by precipitation and snow and ice melt.

Groundwater may be trapped in basins created over geologic time that are not connected to any stream flow, and hence are neither naturally part of the hydrologic cycle nor recharged. Taking water from a basin that is not recharged through the hydrologic cycle is similar to mining a mineral: once it is depleted, it may be gone forever.

Water is substantially different from any other natural resource in that use for one purpose at a given time and location does not necessarily displace its use elsewhere, or at a later time, for the same or another purpose. Water can be used many times as it moves downstream. After water has been used for one purpose, irrigation for example, the same water can return to the stream either as surface or hydrologically connected groundwater. The return flow contributes to recharge the stream. It could conceivably irrigate several crops before it reaches the oceans and begins the hydrologic cycle anew.

Water is abundant when observed on a planetary scale of the earth. However, it can be extremely scarce when considered locally. Weather, geology, and the location and size of landmass exert substantial influence on local temperature and pressure and also on the amount of water vapor contained in the air. These determine in large measure how much rainfall or snowfall will occur in a local area.

Historical weather patterns can give some measure of predictability of rainfall and snowfall, but the annual variability can be significant. All of these circumstances combine to make the geographic area that includes New Mexico, an arid to semi-arid area, which is subject to episodic droughts. There is considerable variation in snowfall and rainfall from one area of the state to another as well as seasonal variations.

The timing of the availability of water is critical to determine whether it may be used beneficially for humankind. A flood can provide a substantial quantity of water, but cannot be used efficiently unless it can be stored and released when needed. For irrigation, water must be delivered at critical times during the growing cycle of the specific crop. Water for municipal uses must be available throughout the year. Location of water is also important. Abundant water may exist in lakes, streams, and underground basins, but if the water cannot be delivered to a place of proposed beneficial use at a reasonable cost, it may worthless for that use.

Reservoirs, canals, and ditches or any other product of human activity do not create any water. However, they do allow control of the timing of the availability of the water and the location where water can be delivered, and hence can materially increase beneficial uses to which the water can be applied. Many beneficial uses would be impossible to accomplish without means to store and transport water from its source to its place of utilization.

**Water Law in New Mexico**

In New Mexico, the surface and groundwater are owned by its people. Individuals, governments, and corporate entities can acquire private property rights to use the waters, if they apply the waters to beneficial uses.

Where there are disputes or matters that need to be resolved about water rights, parties can go to the district courts of the state. The rules of engagement for legal battles for water rights are set out in the New Mexico Constitution; federal and state legislation; federal and state appellate decisions; procedural rules adopted by the Supreme Court; and case management orders entered by district judges.

**The New Mexico Constitution**

> "The unappropriated water of every natural stream, perennial or torrential, within the State of New Mexico, is hereby declared to belong to the public and to be subject to appropriation for beneficial use, in accordance with the laws of the state."

> "Priority of appropriation shall give the better right."

"Beneficial use shall be the basis, the measure and the limit of the right to the use of water."

Types of beneficial uses include irrigation, domestic, livestock, municipal, commercial, and industrial.

The constitutional provisions form the Prior Appropriation Doctrine. This doctrine is included in the laws of many of the arid western states. It has been the law in the state from territorial days before the Constitution was adopted and to a certain degree, even before New Mexico became a part of the United States.

In water cases, like any other matter brought to the courts, litigants are entitled to due process. Due process encompasses the idea that laws and legal proceedings must be fair. Because of the number of claimants, ensuring due process can present the court with major difficulties. The United States Constitution and the New Mexico Constitution guarantee that the government cannot take away a person's basic rights to life, liberty or property without due process of law. A due process issue that frequently arises in adjudications is what can be adequate notice to thousands of water right claimants when the court considers and rules on matters.

## New Mexico Rules of Civil Procedure

The Rules of Civil Procedure are the rules used by courts and parties to provide an orderly determination of procedural questions before a court. These rules apply to adjudications, and are designed to provide a just, speedy, and inexpensive determination of every action. The Rules of Evidence are designed to ensure fairness and to eliminate unjustifiable expense and delay so that the truth and justice prevail. There are five provisional rules of civil procedure that apply specifically to stream adjudications.

Rule 1071.1 NMRA Service and joinder of water rights claimants; responses.
Rule 1071.2 NMRA Stream system issue and expedited *inter se* proceedings.
Rule 1071.3 NMRA Annual joint working session.
Rule 1071.4 NMRA *Ex parte* contacts; general problems of administration.
Rule 1071.5 NMRA Excusal or recusal of a water judge.

All of these have sunset provisions. The Supreme Court considers whether or not to renew them annually.

## The Current New Mexico Water Code

Much water law in New Mexico predates United States sovereignty. Many aspects of prior law have been preserved, but as the sovereigns over the land that became New Mexico changed, the water law changed.

New Mexico's Water Code addresses both surface water and groundwater uses. The New Mexico Territory adopted a Surface Water Code in 1907 that gave the territorial engineer, now the state engineer, supervision over the public's surface water in streams. In 1931, the code was expanded to give the state engineer jurisdiction over groundwater, conditioned on the state engineer declaring an underground basin. The state engineer now supervises the surface water and the hydrologically connected groundwater and manages the stream conjunctively. The term "conjunctively" means that the surface water and the connected groundwater are considered as related parts of the same stream.

Before the New Mexico Water Code of 1907, a surface water right could be established in unappropriated water by building diversion works to use water, that is, damming a stream, constructing canals, and applying water to beneficial use. No permit from territorial (later state) officials was necessary. There was no requirement for a written record of the use to be filed. As a result, evidence to prove beneficial use for pre-1907 rights can sometimes be difficult for pre-water-code claims.

The 1907 Water Code authorized the state engineer to supervise the apportionment to water right claimants of New Mexico's publically owned water. The Water Code modified the requirements for a claimant to perfect the right to use public waters. Under the code, an application to the Office of the State Engineer (OSE) for a permit and the state engineer's issuance of that permit are mandatory precursors to obtaining a surface water right. The application form requires the applicant to state the amount of water and period or periods of annual use, and all other data necessary for the proper description and limitation of the right applied for, and information, maps, field notes, plans and specifications to show the practicability of the construction and the ability of the applicant to complete it.

When an application is filed, the state engineer must determine whether there is unappropriated water. He must also evaluate applications to ensure that the proposed use is an acceptable beneficial use and will not impair existing rights, is not contrary to conservation of water within the state, and is not detrimental to the public welfare. The permit also spells out specific limitations to the use of the water. If granted, the permit authorizes the permitee to construct the diversion works within a specific period of time. Possession of a permit is not evidence that water has been diverted and applied to beneficial use, but merely documents the state engineer's permission to proceed and authorizes the applicant to construct the diversion works.

The claimant's application and the permit issued by the state engineer provide a record of valuable information that is not available for pre-1907 rights. A water right perfected before 1907 is just as valid as one acquired after, but may be more difficult to prove because there was no requirement to file for record evidence of diversion and beneficial use.

There are six elements of a water right: the priority, amount, purpose, periods and place of use, and for irrigation water rights, the specific tracts of land irrigated. These elements are described in the application and included in the permit. Of these elements, the two that result in the most disputes are priority and amount.

Priority is a basic tenet of the Prior Appropriation Doctrine. It is the comparison of the date one water user acquired his or her water right to the date that others obtained their right. The Prior Appropriation Doctrine dictates that the owner of the first water right in time is first in to receive water. In times of scarcity, the claimant with the earliest priority date receives all his or her water first, then the other claimants receive their water in order of priority. The earliest priority date is the most "senior," and water right owners who have earlier dates are referred to as senior. Those who have later dates are referred to as "junior" water right owners. In times of droughts or water shortages, a senior water right owner can make a "call" to enforce his or her priority right to water in the stream system.

The element of "amount" of water can result in much acrimony among water users. For irrigation rights, for example, a claimant cannot establish a water right for amounts in excess of what a reasonably skilled farmer would need to grow crops. Waste is not a beneficial use. Therefore, for irrigation, the amount of water acquired in a water right is determined by the quantity needed to competently grow crops without waste. This is referred to as the consumptive irrigation requirement (CIR). Farm delivery requirement (FDR) is the total water that must be delivered to farms to irrigate the crop, considering unavoidable losses. FDR is the CIR plus farm losses due to evaporation, deep percolation, and other unavoidable surface waste and nonproductive consumption.

Other types of beneficial uses have other standard quantification measures appropriate to the type of use.

It is a common belief among permit holders or their successors in interest that a permit creates a water right for the quantity requested. However, possession of a permit is no evidence that water has been diverted and applied to beneficial use. The permit only authorizes a claimant intending to divert water, to prepare diversion works to divert water to beneficial use, up to a specified quantity, but it does not create a water right. Water users can be confused if the amount of water applied to beneficial use is less than the amount in the permit. The amount of the water right is the amount that a user puts to beneficial use and not the amount described in the permit.

The 1931 Groundwater Code, unlike the 1907 Surface Water Code, does not automatically give the state engineer jurisdiction over groundwater. The state engineer cannot require permits for new groundwater uses until the affected groundwater basin has been declared, that is, the boundaries of the basin are fixed by identifying and declaring them as reasonably ascertainable. After the state engineer declares such a basin, he acquires jurisdiction of the groundwater. The state engineer has recently identified and declared all underground basins in New Mexico that had not previously been declared. Thus, anyone seeking to acquire a new groundwater right in New Mexico now must obtain a permit from the Office of the State Engineer (OSE) to drill a well.

Before 1907 for surface water rights, and before the state engineer's declaration of an underground water basin for groundwater, if someone disputed another's claim, the challenged owner had to prove when, where, and how much water he or she diverted to beneficial use to establish the water right.

After the dates when permits are required, upon application, the state engineer must first determine whether there is unappropriated water. The permit specifies the amount of time a claimant has to complete the diversion works. The permit holder subsequently acquires a water right only as to the amount timely applied to beneficial use.

The state engineer must have certain information regarding the water usage on the stream to adequately supervise the apportionment of the public waters. Permits alone do not give the state engineer adequate information. That information may either come from "licenses" issued by the state engineer or the court's adjudication of water rights. The Water Code sets out specific steps that the state engineer should take after issuing the permit. The step that should be followed after issuance of a permit is the OSE's timely inspection and issuance of a license after the permitee has completed the diversion works. The license will state the specific amount of water that can be diverted to beneficial use and establishes an administrative record of the water being diverted. The amount of water described in the license is the amount of the actual application to beneficial use. It can be less than the amount stated in the permit; it cannot be more.

Where a license exists, the OSE has made a field check to verify and measure the application of water to beneficial use, and the license provides substantial evidence of the water right and reflects agreement between the state and the water right licensee. The court, in a subsequent adjudication, need only make determinations about changes that may have occurred to the water right since the issuance of the license and conduct an *inter se* proceeding before entering a final decree. Historically, the OSE has not issued many licenses.

Since all waters in a stream system are hydrologically connected, use by one water right owner can affect the right of other water rights owners. As a consequence, water users on a stream system have a due process right to contest other water users rights if they so choose. Typically, this is done after an owner's sub-file matters have been resolved with the OSE. These proceedings are referred to as *inter se* matters and are explained more thoroughly in the section on stream adjudication.

## Federal Law

Preemption
The Supremacy Clause of the United States Constitution states that the "Constitution and the laws of the United States . . . shall be the supreme law of the land . . . anything in the constitutions or laws of any state to the contrary notwithstanding." This is referred to as the Federal Preemption Doctrine. In water law, there are several federal statutes that preempt state law. The Endangered Species Act and the Clean Water Act are examples. The Endangered Species Act, in particular, has generated heated legal battles because it can have a direct impact by limiting otherwise valid water rights.

Reserved Water Rights
The relationship between the United States and Native peoples residing within its boundaries has been filled with strife and misunderstanding. In the

preface to his book, *Indian Wars of the West*, Paul I. Wellman, describes that struggle.

> In the Old West the beginnings of the Machine Age encountered the last vestiges of the Stone Age. Between these two extremes of human culture there was no common ground. The Indian could not understand the paleface's land hunger. To him the earth and its creatures belonged to all, the free gift of the Great Mystery. That one should build a fence around a little corner and say, "This is mine," was repugnant. The white man possessed the repeating rifle, the telegraph and the railroad. The Indian had only his primitive weapons and his native courage.
>
> Remorselessly, the Machine Age engulfed the Wilderness. The white man made solemn agreements that were not kept because the Senate of the United States had a habit of never getting around to ratifying the treaties that military leaders and peace commissioners signed with the Indians. Nearly all their land had been taken from them except that which appeared worthless.

This clash of cultures in the United States between the indigenous peoples and the European immigrants, in many cases, led to military wars that were not ended until the warring parties signed treaties providing for reservations of public lands for the purpose of developing Native American agrarian communities. Resolution of disputes between the United States and native tribes, either by actual war or negotiated agreement resulted in a major impact on water law throughout states that had Native American reservations of land.

When the United States set aside reservation lands by treaty and moved Native Americans onto them, it did not address the question of water rights.

In the 1908 case of *Winters v. United States*, the U.S. Supreme Court interpreted the intent of the treaties regarding water, and established the federal reserved water doctrine. In the *Winters* case, the U.S. Supreme Court found that water rights were implicitly reserved for future use in an amount necessary to fulfill the agrarian purpose of the reservation. This decision is the basis of the court decisions regarding water rights on Indian reservations today. The United States, as trustee, holds legal title to reserved water rights on Indian reservations.

The *Winters Doctrine* is also the basis for other types of federal reservations for other purposes.

Federal law-based reserved water rights, just like state law-based water rights, are bound by the basic tenant of the Prior Appropriation Doctrine, that is, first in time is first in right. If a priority call is made, holders of federal reserved rights take in order of their priority. Nevertheless, there is a significant difference in how federal reserved water rights are quantified and maintained in comparison to state law-based water rights. In state law-based water rights, an owner must put the water to beneficial use within a reasonable time and may be subject to forfeiture or abandonment if not used in the future. There is no time limit for a claimant of a federal law-based water right to appropriate for beneficial use.

This presents difficult legal and factual problems, and can generate heated fights in adjudications. A federal reserved water right typically has an early priority date. An owner of state law-based water rights, which are junior to a federal reserved water right, may reliably have received water for decades before a senior federal law-based water right owner makes a priority call. Even if the senior federal law-based water right owner has never made a call since the date of reservation, or has never used all the water for which he is entitled, the senior federal reserved right may have a legitimate claim to most or all available water in the stream. Thus a junior water right holder might be shut out from receiving water when a priority call is enforced.

**Stream Adjudications**

Adjudications are the responsibility of the courts and are required under the Water Code. The federal district courts and the state district courts have concurrent jurisdiction. This means that an adjudication can be brought in either court. If the United States claims some right in the stream system, it is typically joined as a party. All governmental entities from the federal government on down are protected from being sued, unless there has been express waiver of immunity. The United States waived governmental immunity for stream adjudications when Congress adopted the McCarran Amendment to the federal Reclamation Act, and consented that it could be joined as a defendant in state courts. The Amendment required that the adjudication encompass all or a comprehensive part of a stream system.

The Water Code directs the state engineer to make hydrographic surveys and investigations that record existing water usage for each stream system in the state that can record existing water usage. Thereafter, the code directs the state to enter an adjudication suit in district court for the determination of all rights to the use of such water. All who claim the right to use of the stream system's waters must be joined as parties. Water adjudications determine if there is unappropriated water in the stream. They further determine ownership of each water right and create for the state engineer an inventory of all active water uses within a stream system, and thereby provide the state engineer with sufficient information for him to supervise the public waters.

Users of water, that may be involved in adjudications, include Indian tribes and pueblos, United States agencies, the State of New Mexico, counties, irrigation districts, conservancy districts, municipalities, corporate entities, acequias, and individual users.

An acequia is a community-operated ditch system used for irrigation. The legal status of New Mexico acequias is a legacy of the laws of Spain and Mexico, which had been adapted from Arabic law. Many acequias were established from the time of Spanish occupation of what is now New Mexico. Acequias continue to this day to provide water for agriculture in parts of New Mexico, and state law sets out responsibilities for users of waters from acequias. Although some may be modern structures, the majority of acequias are simple, open ditches with dirt banks.

Historically, the state engineer prosecutes stream adjudications in three phases. First the state engineer conducts a hydrographic survey, after which the Attorney General appoints the legal staff of the OSE as deputy attorneys general. The staff file an adjudication and initiate a sub-file phase. Next is the final *inter se* phase. In the hydrographic survey phase, the state engineer inventories all possible water rights in a stream system by examining public records and conduction of field investigations. In the sub-file phase, the state joins the identified water users to the case, using the hydrographic survey report as a guide. The state makes offers to each individual claimant based on the information in the hydrographic survey report. The matter is resolved by a sub-file order frequently stipulated between the water right claimant and the state. If agreement cannot be reached sub-file matters are resolved by trial before a judge or special master.

The distinction between sub-file proceedings and *inter se* proceedings arose because appellate courts have approved procedures to separate issues or "bifurcate" stream adjudications into the two proceedings. A person's water right claim, in a legal sense, is adverse to the state and potentially adverse to all other water right claimants on the stream system. If a dispute is settled between the water right owner and the state, and other water right claimants want to object to the sub-file order, an "*inter se*" proceeding is commenced.

Ordinarily, sub-file proceedings for all claimants are resolved before *inter se* proceedings are commenced. However, appellate courts have also approved an "expedited *inter se*" procedure to permit resolution of issues of a claimant's assertion of a water right, both with the state and all disputing parties, before other claimants' sub-files have been resolved.

Completed adjudications will give the state engineer the fundamental information necessary for him to supervise our public waters. They will materially reduce the possibility of New Mexico being sued by another state or other sovereigns for their equitable share or treaty share. They will reduce uncertainty of ownership, priority, and quantity and other elements of a water right. Water right owners who want to sell their rights and purchasers of those rights will have substantially better information that should simplify the market in water rights.

The adjudication court faces many challenges. A stream adjudication is an adversarial process. Water disputes in adjudications can be fierce and heated. They can be very expensive to prosecute and can take decades to resolve.

The court is directed by rules to use speedy and inexpensive procedure and hence is confronted with the need to investigate and implement procedures to shorten the time to complete the adjudication. In stream adjudications in New Mexico, the goal of fairness and justice is not materially different from any other civil case. However, the number of claimants, particularly those not represented by counsel and the number of water rights in an adjudication can present serious obstacles to attain the goals of avoidance of delay and unnecessary expense. There is a dichotomy presented. Without careful attention by the court, speedy and inexpensive procedure may result in injustice. Alternatively, the need to ensure justice can slow down the pace and increase the expense of an adjudication.

**Lower Rio Grande Adjudication**

The Beginning

The Lower Rio Grande Stream Adjudication was initiated on September

12, 1986 by Elephant Butte Irrigation District (EBID). The irrigation district asked the District Court to order the state engineer to conduct a hydrographic survey and conduct a general stream adjudication of a portion of the Rio Grande in southern New Mexico. Although there was initial confusion whether the Elephant Butte Reservoir was to be included, the segment of the Rio Grande now encompassed in the Lower Rio Grande Adjudication is from the Elephant Butte Dam south to the Texas state line and the international border with Mexico.

For the first ten years, the court addressed various legal issues. Initially, the state engineer vigorously opposed completing a hydrographic survey or initiating a stream adjudication. It argued that the state engineer did not have the financial and staff resources to conduct an adjudication. The state and other parties questioned whether the court could order a hydrographic survey and compel the state to undertake an adjudication. The court determined it had jurisdiction to order the state to undertake a hydrographic survey.

Certain defendants filed a petition to move the adjudication to federal court in 1987. In 1989, the federal court denied the petition and returned the adjudication to state court.

The description of the stream to be adjudicated did not include the northern two thirds of the Rio Grande in New Mexico from the Colorado state line to the Elephant Butte Dam. There was a question whether the Lower Rio Grande could qualify as a comprehensive stream adjudication as required by the McCarran Amendment of the Reclamation Act. The Rio Grande Compact requires New Mexico and Colorado to deliver a determinable quantity of water to the Elephant Butte Reservoir for the benefit of southern New Mexico and Texas. In 1993, the Court of Appeals decided that the Rio Grande Compact obligations qualified the Lower Rio Grande Adjudication as a comprehensive McCarran stream; that the United States had waived immunity; and that a comprehensive adjudication could proceed in state court.

There was a second attempt to move the adjudication to federal court. The New Mexico Federal District Court retained jurisdiction, but remanded the adjudication to the state court.

The adjudication process, as set out in the Water Code, began in earnest in 1995, when the Court ordered the state to complete a hydrographic survey and start an adjudication. The state dropped its objection to commencing the adjudication, obtained funds from the legislature, and contracted with a private contractor to do a hydrographic survey. Actual adjudication of water rights began on May 24, 1999, which was the date that the state submitted the first sub-file order to the court to approve and enter.

Staffing

Since 1986, when EBID filed and requested the court to order the state to begin the adjudication, there have been four presiding judges, soon to be five, and the state has been represented by twenty-one attorneys. It is likely that these numbers will continue to increase until the adjudication is complete. Because of the turnover, historical memory can be lacking. The numbers of judges and attorneys for the state involved in adjudications over a considerable span of time can cause inconsistency in its prosecution within a specific adjudication.

The OSE divides its attorneys into adjudication teams for each stream being adjudicated. As a result, although there is some coordination, an adjudication on one stream system may proceed somewhat in isolation from other adjudications. Until recently, each adjudication either had a sitting judge, who was assigned a regular docket in addition to the adjudication or a judge *pro tempore* who presided only over one adjudication. A *pro tempore* judge is authorized by the New Mexico Constitution if a district court is unable to expeditiously dispose of a case or cases in the district. Presently, the Supreme Court has assigned one judge to four adjudications.

In limited circumstances, special masters are authorized by the Rules of Procedure, and appointed to assist with an adjudication.

Claimants

All who claim water rights must be made parties. To date, nearly 18,000 claimants have been joined to the Lower Rio Grande Adjudication. In nearly all adjudications, the state is the prosecuting party. Except for the state, different adjudications will likely have different parties, depending on who claims rights to use water in the particular stream system.

There are three distinct groups of water right claimants in stream adjudications that raise different due process considerations.

First, there is a group that includes major claimants who frequently and fiercely litigate water issues. They are well represented by counsel competent in water law. In the Lower Rio Grande Adjudication, this group includes the State of New Mexico, Elephant Butte Irrigation District, the United States, the City of Las Cruces, the City of El Paso, Stahmann Farms, New Mexico State University, and El Paso County Water Improvement District No. 1

(EPCWID#1). Recently, corporate entities have been formed that participate in the adjudication on behalf of their members/claimants and are also well represented. These entities include the New Mexico Pecan Growers' Association and the Southern Rio Grande Diversified Crop Farmers' Association. They may want to participate in all major issues before the court, or they may only wish to participate when certain specific matters are at issue. Because litigants in this category are well represented, the legal interests of these claimants are adequately advocated.

Secondly, there is a group that includes claimants who are represented by counsel with limited experience in water law and procedure. Because of the arcane aspects of water law, their lawyers may need to devote time to research and to prepare adequately. This can be costly for claimants who may not, in fact, have any significant dispute. Because of the length of time matters are pending before a specific water right can be addressed by the court, legal fees can add up to significant amounts. Like claimants in the unrepresented group, a majority will settle issues related to their claim without a significant amount of litigation.

Lastly, there is a group that includes large numbers of water right claimants, who are not represented by counsel and are primarily concerned with receiving irrigation water in a timely fashion, or a guarantee of their use of domestic wells. Some may be sophisticated in water law and are capable of asserting their rights adequately. Others in this group may not understand important legal aspects of water law. Most stream system issues that may affect their water rights may be presented adequately by attorneys representing other parties, but this is not a certainty. Many in this group do not trust in the *bona fides* of the Office of the State Engineer and wrongly believe that they are being sued so that the state can take away or limit their water rights.

This group can present a problem of delay to the court because many may lack an understanding of legal rules and sometimes do not follow procedures adequately to protect their claims. On analysis, many of these claims will not have factual issues in dispute and hence should be resolved efficiently by settlement or mediation. Even when factual disputes are identified, the actual number of claims requiring a trial to resolve disputes is likely to be limited, once claimants have an adequate understanding of water law and the process of adjudication.

The Joe M. Stell Ombudsman Program was started to provide a source of this information. With adequate understanding of water law and adjudication procedure, unrepresented parties can utilize negotiation with the state or mediation to reach a stipulation with the state, or, if necessary, be prepared to litigate. Those cases that cannot settle are resolved by trial in front of a special master or a judge.

EPCWID#1 is a special case. It filed a Motion to Intervene. There was concern regarding the jurisdiction of a New Mexico state court over a Texas irrigation district. The motion to intervene was stayed. However, the court granted *amicus curiae* status. *Amicus* participation by EPCWID#1 provides the court with a perspective and a source of information that might not otherwise be available.

Docket Number/Sub-file Number/Case Number/Stream System Issue Number

The docket number of the Lower Rio Grande Adjudication originally was CV 86-848. When a civil case is filed with a district court, it is assigned a number. CV denotes that the case is a civil case. The first two digits (86) indicate that the case was filed in 1986. The 848 signifies that it was the eight hundred forty eighth case filed in 1986.

In 1996, the New Mexico Judiciary rolled out a new case management system called FACTS for all matters filed in court. To minimize the resources required to convert the old system to FACTS, the Lower Rio Grande Stream Adjudication was given a new docket number in the new software. It is now CV 96-888. The earlier part, CV 86-848, was consolidated into it. The change in docket numbers may be confusing to those who are acquainted with the typical court docketing system.

The state engineer's legal staff organizes its water right information in sub-files. Each water right or a related group of water rights are assigned a sub-file number. In the Lower Rio Grande Adjudication, a case number is assigned to each claimant to the sub-file water right. The state's August, 2010 status report indicates there are 13,717 sub-files and 17,830 claimants in the Lower Rio Grande Adjudication.

Stream system issue proceedings are explained more thoroughly in the next paragraph below. They are numbered SS 97-101, *et seq*. The court gives them a stream system issue number after an issue has been designated as a stream system issue.

Case Management Orders

Ordinarily, the presiding judge and the major parties develop case management orders for their particular adjudication. Case management orders may differ in material aspects from adjudication to adjudication. This has both positive and negative consequences. They have the potential of causing lack of consistency among adjudications, but particularized orders also permit addressing issues that are specific to an adjudication. Case management orders are developed with input from parties. Specialized adjudication rules could address the procedural inconsistencies among adjudications, but except for five provisional rules, the Supreme Court has not yet adopted comprehensive adjudication rules.

Sixth Amended Order Filed September 14, 2009

The first case management, entered by the court in the Lower Rio Grande Adjudication, set out sub-file procedures. A seventh iteration of the original case management order is now in effect. The procedure is designed to ease the uncertainty and concern of claimants who do not have attorneys to assert their individual claims. It gives individual claimants the option to combine with other claimants to minimize costs. The Sixth Amended Order requires the state to notify the ombudsman program and for the ombudsman program to contact claimants to provide them with information important to the assertion of the claimants' rights. It provides for simplified forms to respond to the service of the complaint, and explains the consequences of failing to respond.

This order covers both sub-file and *inter se* proceedings and controls when specific water rights are to be determined. The order begins with definitions of terms. It defines stream system issues, *inter se* proceedings and expedited *inter se* proceedings. The Rules of Civil Procedure apply except as expressly modified.

The Sixth Amended Order provides for simplified forms for water right claimants to use when served with a summons and complaint. It provides for coordination with the ombudsman program. In compliance with the order, the state must make offers of judgment that are the state's proposal to stipulate to claimants water rights. The order gives directions to claimants regarding objections to the offer of judgment. It explains the stipulated sub-file orders, sub-file orders-implied consent and sub-file orders-default.

The order explains that claimants can negotiate with the state to resolve their water rights by stipulation, and directs the parties to mediate through the court-annexed mediation program if initial negotiations are unsatisfactory. If the claimant cannot resolve issues with the state by stipulation, the order provides that the matter will be tried before a special master or judge. It also provides that stipulated sub-file orders, implied consent sub-files and default sub-files are not appealable or modifiable except as permitted under Rule

1-060 (b) NMRA, or as may be necessary after *inter se* issues are decided.

The order further provides that several parties may be represented by one attorney if there is no conflict of interest; corporate entities may answer and file updates of their address and ownership records without an attorney and, when a corporate entity wants the court to take action or grant relief, it must retain an attorney. Individual claimants may form an independent, non-governmental, voluntary, corporation or other appropriate corporate entities to act on behalf of its members to resolve issues between its members and the state. There must be written confirmation that its members have authorized the corporate entity to act on their behalf.

First Amended Case Management Order Authorizing Notice by a Monthly Report, filed September 14, 2009

The court has entered a case management order addressing service of process. This case management order is on its second iteration. The ordinary rules of civil procedure require service by first class mailing after parties have been joined. When motions on stream system issues are filed, the cost of mailing notice to all claimants would be high. The state engineer has identified the names and addresses of almost all of the water right claimants in the hydrographic survey. First class mail should be sufficient for due process.

This order provides for notice to claimants through a Quarterly Report for matters of general concern to the adjudication, stream system issue proceedings, and expedited *inter se* proceedings. The order explains how a stream system issue or expedited *inter se* proceeding may be initiated. The quarterly reports are posted on the New Mexico Judiciary's website, www.nmcourts.gov (Click on Lower Rio Grande Adjudication.). The posting of the Quarterly Report and the posting of documents on the website is effective service on all claimants.

Parties must file timely notices of intent to participate in stream system issue proceedings. Lists of parties with their addresses, who have filed notices of intent to participate, are published on the website. The ordinary rules of civil service of documents apply to parties participating in a stream system issue proceeding. A final decision by the court on a stream system issue, or in an expedited *inter se* proceeding, will bind all parties whether or not they have participated in the proceeding.

To date, four stream system issue proceedings have commenced. There are approximately 30 parties participating in each of the following stream system issue proceedings.

SS 97-101: Consumptive Irrigation and Farm Delivery Requirements for All Crops in the Lower Rio Grande Basin.
SS 97-102: Elephant Butte Irrigation District's Claim to Underground Waters on 90,640 Acres of Its Members' Lands.
SS 97-103: Priority, Transferability, and Beneficial Use Elements of a Domestic Well Water Right.
SS 97-104: The United States Interests in the Stream System.

SS 97-101 has been set for trial June 6, 2011. The Court has recently received notice that SS 97-102 has been resolved by stipulation. Scheduling deadlines are currently being considered in SS 97-103. SS 97-104 has been partially stayed pending mediation.

The order provides an opportunity for all claimants to participate in stream system issue proceedings, but has the practical effect of reducing the number who will actually participate to those represented by knowledgeable attorneys, or parties who are familiar with rules of litigation. Participating parties must follow the rules of civil procedure with respect to other participating parties. The order provides an inexpensive method of giving notice to claimants who are not participating parties by posting activity on the judiciary's website. This protects the due process rights of those who choose not to participate and will greatly reduce the cost of service, and will allow the court to ensure that stream system issues are resolved promptly.

Order for a Hydrology Committee

In 1999, the state, EBID, the United States, the City of Las Cruces, the City of El Paso, New Mexico State University, joined by Stahmann Farms, Inc. and *Amicus Curiae* El Paso County Water Improvement District No. 1 established a Hydrology Committee. The purpose of the committee was to promote cooperation among the parties and their experts and to provide technical assistance to the parties. The protocol expressly provided that the Hydrology Committee would not act as a technical advisor to the court.

The court has recently entered an Order for Hydrology Committee that materially changed the function of the committee. The committee will now operate in a manner similar to a court expert as described in Evidence Rule 11-706, NMRA. The changes to the Hydrology Committee were based on procedure and rules adopted by the Colorado Supreme Court. Any party may name, but is not required to name, a hydrologist to the committee.

The members of the Hydrology Committee must disclose their expert reports to each other and discuss the matters of fact and expert opinions. Thereafter, they jointly submit to the presiding judge a written statement setting forth the disputed matters of fact and expert opinion that remain for trial. No claimant is required to name an expert to the committee. Any claimant may retain an expert, who need not be a member of the hydrology committee, to testify at trial.

The Hydrology Committee should narrow the issues that need to be addressed by the court. This order encourages parties' experts to have open discussion on matters that require the expertise of hydrologists and to advise and explain hydrology issues to the court that are actually disputed. This should reduce the overall expense litigating complex hydrological questions.

**The Water Wars Go On**

When the courts of New Mexico conclude all adjudications and determine and inventory all the water rights in New Mexico's stream systems, the legal battle for water rights will not end. Future disputes may be as contentious as past or present ones, and it is likely that the judiciary will be involved in resolving water disputes. The past has shown that the law has been resilient to meet changed conditions, and it is inevitable that water law will change to meet changing societal conditions in the future and the legal battles over water will continue. The importance of water to our state is self evident. Contentious water disputes have been with us for a long time and will continue.

The well of a courtroom is a safer place to fight water wars than the battlefields of Sumeria, but sparks still fly. The following are edited versions of statements from briefs recently filed in the Lower Rio Grande Adjudication.

> "Rather than address the ... listed facts as required by the rule, [Party B] mis-characterizes [Party A's] motion in a nonsensical fashion and then wastes space arguing against [Party B's] own mis-characterizations. . . . This characterization of [Party A's] position is wholly false, unsupported and unsupportable."

> "As a legal matter, [Party C's] argument is absurd."

> "Both [Party D] and [Party E], in keeping with a litigation strategy that was always unfounded and been repeatedly rejected by this court, continue to attempt to confuse [issues]. . . ."
>
> "A stay . . . makes about as much sense as turning off the microscope lamp just when you place the specimen under the lens to view it. Litigation illuminates. Cross-examination clarifies the contours of amorphous claims; and, as every litigator knows, it powerfully motivates real settlement discussion. If this adjudication is to resolve within our joint lifetimes, litigation progress needs to be encouraged, not stymied."
>
> "[Party F] has made extensive, confusing and contradictory claims . . . in this litigation, and the nature and scope of [Party F's] claims . . . bear directly on the [water right claims] held in whole or in part by [Party G] . . . [Party F] appears to be pursuing a strategy of obfuscating its claims."

The weapons of litigation are words and arguments. Motions are a strategy of attack against opposing parties. Adversaries lob sharp invectives at each other like cannonballs in a real war. Nevertheless, when compared to real war, litigation is a gentle kind of battle, restricted by legislation, by rules, and by court orders. Adversarial clashes that sometimes flare in adjudications can be ameliorated by a judge's careful and firm management.

## Conclusion

> Water is the subject of many musings:
>> "When the well is dry, we know the worth of water."
>> "Water flows uphill towards money."
>> "Water is necessary for life on Earth."
>> "All the water that will ever be is, right now."
>> "Water is worth more than oil."
>> "Whiskey's for drinking, water's for fighting."

It is readily apparent from these quotes that water is of utmost importance to humankind, and, by implication, contentiousness is endemic in human resolution of water issues.

A judge is responsible for managing his or her case load. Within the stricture of legislation, court rules, and resources, the court has utilized its authority to try to devise more efficient procedures. The case management orders are works in progress. These procedures have been developed by trial and error. Hindsight has identified both mistakes and successes. There is no guarantee that these current procedures will lessen the time to complete adjudications, but there is a reasonable chance that they will.

The substantive and procedural difficulties inherent in the New Mexico Water Code have been obstacles to the just, speedy, and inexpensive determination of the adjudication. Justice can be impaired by the cost of litigation. It is an axiom that justice delayed can result in justice denied. Stream adjudications can take decades to resolve at considerable expense to the state and claimants. The art of litigation is difficult, water law arcane, and ordinary rules of procedure not always easy to apply fairly under circumstances that arise in adjudications.

The goal of a just, speedy, and inexpensive determination of water rights in adjudications remains elusive, but the search continues for improved rules and procedure to reduce the contentiousness in the continuing legal combat over water rights.

# 3

# Ready to Fight: Steve Reynolds—Institution, Engineer, Litigator

## John W. Hernandez

**At Statehood, Were the People of New Mexico the Problem?**

Maybe. You might make the case that our relatively recent background as pioneers was responsible for at least a part of our proclivity to "fight" over water in those early years. Is it fair to say that we are different and more likely to be involved in a fight over the right to use water, than are our other western neighbors? Arizona, Colorado, and Texas don't seem to have their supreme courts involved in as many water conflicts as we do.

And what state, other than a "ready to fight over water" bunch would include the heart of their water code in the state constitution? Other states may have done it, too, but our grandfathers thought it important enough to put the core of our water law in our 1912 constitution:

> Article XVI Irrigation and Water Rights
>
> > Section 1. All existing rights to the use of any waters in this state for any useful or beneficial purpose are hereby recognized and confirmed.
> > Section 2. The unappropriated water of every natural stream, perennial or torrential, is hereby declared to belong to the public and be subject to appropriation for beneficial use. . . . Priority of appropriation shall give the better right.

> Section 3. Beneficial use shall be the basis, the measure and the limit to the right to the use of water.

Water wars are still going on in New Mexico even though the words in the state constitution seem straightforward and easy to understand. But the record says something different. After 90 years of statehood, the state's higher courts had already rendered 160 opinions in water cases, almost two a year. That's just the state court cases; interstate conflicts go to the federal courts where we also have had more than our share. Some federal water rights adjudication cases in New Mexico have gone on and on. These cases did not just involve legal conflicts among water users; other factors seemed to have been involved. The folks that came to New Mexico were by nature a scrappy bunch, particularly those charged with the protection of the little water that we do have.

Maybe it wasn't our general citizenry that was fervent for a fight, but it may have been our chief water administrators, the state engineers. Looking over the record of some of our state engineers could easily lead to the conclusion that our "water mayordomos" were a major part of the problem. All of our state engineers had cases that went to state or federal courts for resolution, but there was one that beat them all: Steve Reynolds.

## Steve Reynolds, State Engineer 1956–1990

In the sections that follow, I will talk about just one of our state engineers—Steve Reynolds. With few exceptions, all of us who have authored sections for this book worked for or knew Steve Reynolds firsthand. It would be bizarre for us not to tell you about him, and in particular his personal involvement in our many legal battles during his 34 years of service. Reynolds and his gang fought off federal Forest Service claims to New Mexico's water and they won victories in the New Mexico Supreme Court. You will find stories about these wars and others in this book.

My story about Steve Reynolds is divided into three sub-sections: Reynolds the Institution, Reynolds the Engineer, and finally, Reynolds the Litigator. It is the combination of these three distinguishing personal abilities, acting in concert, at all times, that made Reynolds a great water administrator and a water warrior.

## Reynolds the Institution

Steve Reynolds was an institution. Thirty-four years of water rule is a long time. He was broadly known and considered to be more powerful than many of the governors he served under. Reynolds was widely respected for many reasons. His alleged power was certainly one reason. He always denied that he was a politically powerful person and on a couple of occasions said:

> "Some may logically conclude that the Office of the State Engineer is vested with great power. As are many logical conclusions, that one is not well founded. Our statutes on water fill a volume about two inches thick and the case law—our Supreme Court opinions stack up another 3.5 inches on 8.5 x 14 inch sheets. The statutory and case law rather effectively constrains the discretion of the State Engineer. Furthermore, the State Engineer's every decision, act, or refusal to act is subject to appeal to the District Court; and this recourse is not infrequently sought."

Reynolds gave hundreds of speeches to many different organizations, and when he wanted to send a message to some group, I believe that he deliberately sought an opportunity to let them know what was on his mind. And he did tell them!

His speeches sometimes showed those aspects of his character that help the reader to understand his strong sense of purpose. For example, in one speech he talked about the ideas of an author on "megatrends" in our society. Steve quotes the author as saying "trends, like horses are easier to ride in the direction they are already going." Reynolds then put forth his notions on moving society in his direction when he said, "The author may know about megatrends, but I'm not sure he understands horses. Upon getting into the saddle—right then, you have to let the horse know who is going to direct its course. If we all just ride along, we may have an unhappy destination." Reynolds certainly didn't "just ride along." He was a determined leader who set the direction for water management for a third of a century, and even now his past actions point toward the road that current water decision makers should take.

Reynolds got along well with most of those in power in New Mexico and the U.S. He had good rapport with New Mexico legislators and he often spoke of the wise leadership offered by this legislator or that. He always spoke about state

Senators Ike Smalley and Aubrey Dunn and of their help. In a speech, Reynolds quoted Aubrey Dunn as saying, "SER was a great man, but power shouldn't be given to one man." Steve probably smiled at Senator Dunn's comment as there were many who decried Dunn's use of the power of his legislative position. As noted before, Steve never claimed that his office gave him great power.

Reynolds always knew what he was talking about when he went to see a legislator; more important, he understood their interests. He ran the Office of the State Engineer on the cheap, never asking the legislature to spend a ton of money except when our water was threatened by the federal government or by another state. Except for the tail end of his 34 years in office, Reynolds had one of the best, if not the best, assembly of lawyers and engineers in state government. No, he did have the *best* staff around. But it wasn't all peaches and cream at the Office of the State Engineer. Once when I went to work for Steve for a year on sabbatical he said, "John, there's a lot of turf around this place." Yes, there was a lot of internal battling at the OSE.

The U.S. Congress certainly listened to him and that was often. Congressmen also joshed with him. Mo Udall of Arizona once said that Reynolds was a contemporary of President Millard Fillmore and that he was New Mexico's only state engineer. He had the strong support of U.S. senators like Clinton P. Anderson, Ed Mechem, and Dennis Chaves. He once praised Senator Anderson saying, "He is a master of political science with only slightly less acumen in the physical sciences." He was helped greatly by New Mexico congressmen like Manuel Lujan, Tom Morris, Joe Skeen, and others.

I'm sure that some of the governors he served under didn't always appreciate his reputed power, but he was well respected by all. In his speeches, he always had good things to say about governors Ed Mechem and Bruce King. He frequently quoted Mechem in his talks by noting that "Governor Mechem advised me that water as the protracted subject of discussion could become extremely dry," and Steve would then give a talk about water that was far from dry. Reynolds also told the story about freshman Bruce King being the "water boy" for the University of New Mexico football team when Reynolds was an assistant coach, and how, when King became governor, Bruce would call Reynolds his "water boy."

Reynolds did well with the judicial branch of government too, but he always gave credit for his success in court to his great legal staff, particularly Richard Simms. Reynolds would say, "My lawyers tell me what the law requires me to do, and then lawyers in black robes tell me what I did wrong." He often had "good words" to say about New Mexico Supreme Court Justice Irwin S. Moise. For example, Reynolds once said, "Judge Moise is very scholarly, yet highly practical, and that characterizes all of his work." It is interesting to note, that while a member of the New Mexico Supreme Court, Justice Moise supported Steve's right to tie the regulation of groundwater to that of surface water. Moise also supported Steve's contention that the City of Albuquerque did not enjoy the power claimed under the so-called Pueblo Water Rights Doctrine.

Steve was often complementary of other engineers. When the head of the Bureau of Reclamation office in Albuquerque retired, Reynolds spoke at a City Council session to thank Ralph Charles when he said, "As cannot be said of all bureaucrats, Ralph is a gentleman, a scholar, and always pleasant to work with." He recognized the actions of the International Boundary and Water Commission, noting the "good works" of U.S. Commissioner Naren Gunaji and Mexico's engineers. Still, I never heard him say anything really demeaning about anyone.

Reynolds was a good speaker and layered his talks with jokes about various things. He would open with a comment like "First the 'Good News.' Not much has happened in water management in New Mexico in the past 25 years; the bad news is that you will have to listen to me for 30 minutes." When talking to some visiting group in Santa Fe, he was likely to say that the town "was laid out in 1610 by a blind surveyor and a burro." Unfortunately, "Our engineers have not been able to do anything to improve traffic flow. There are few parking spaces in Santa Fe." So, ". . . follow the usual custom of simply abandoning your car when you get near your destination." Two of his favorite comments were: "water runs uphill to money," and "I have learned that for every complex issue there is a simple answer—and it is wrong."

Reynolds was a realist, but he once told of the difference between a pessimist and an optimist. Reynolds said, "The optimist says this is the best of all possible worlds; the pessimist agrees." He had a plaque in his office that read, "If you can keep your head when all those about you are losing theirs, you probably don't understand the situation."

Reynolds often revealed himself as an assured individual through his willingness to poke fun at himself in his speeches. In more than one speech he told the story about "Sheep Hays" when he was the Land Commissioner in the 1960s. Reynolds story went, "We engaged in a bitter controversy in the Portales area over the creation of an irrigation district to appropriate groundwater from beneath the sand hills. 'Sheep' called me after he had been in the area to discuss matters involving state lands. He told me that there were some people in the Portales area that would like to build a monument to me, but that there were a hell of a lot more who couldn't wait."

In his speeches, Steve often called himself a "bureaucrat engineer," or a "hydropolitician," or he would say, "I'm still engaged in on-the-job training in water resources." I'll never forget what he said at the staff celebration of his 30th year as the state engineer. Steve held up a piece of cake and said, "I want to thank all of you, but you should not think of today as the last day of my 30 years of service, but as the first day of the next 30 years." Many of us believed that he, in fact, meant to try.

Reynolds often told stories about lawyers and vice versa. He told of one of his lawyers asking, "What do you call an engineer with half a brain?" Reynolds answered, "What?" His lawyer replied, "Gifted!" We engineers had a harder time kidding him, but I remember going into the office once and finding Reynolds, an arm in a sling, with engineer Carl Slingerland. I asked what had happened and Carl said, "He fell out of a tree; he was raking leaves." Reynolds used to kid me about being an "Aggie" or a "cowboy" because I taught at New Mexico State University, forgetting that I was a University of New Mexico engineering graduate just like himself. On two or three occasions Steve asked me, "When are you going to leave that easy teaching life and come and do some real work?" I did, but only during sabbatical leaves, and then, Steve would mark-up my stuff just like he did everyone else's. He was a better writer than the rest of us engineers. Steve read every letter that went out of the office and marked it up for re-writing. He took his briefcase home every night and worked harder than any of the rest of us. These were some of the characteristics that made him a legend—an institution.

**Reynolds the Engineer Scientist**

I appended the word "scientist" to "engineer." Steve was both a quality, competent engineer, and a knowledgeable applied scientist. He played an important role in a number of scientific areas that might have provided additional wet-water for use in New Mexico. Some of these innovative ideas were: water salvage, rain making, snowpack augmentation, evaporation suppression, desalination, recovery of water from enclosed basins, and importation of water from the Mississippi River. A few water expansion plans pre-dated Reynolds and were successfully implemented. The most significant of these was the San

Juan-Chama Project that was conceived in the early 1920s, but was finally put into effect during Reynolds's watch.

Reynolds was skeptical about some of the federal programs designed to increase western water supplies. He once complained that "the gestation period for major water projects is several decades" and that federal water projects should be subject to state water law. Steve was also a strong advocated for significant federal matching money for major water projects.

Generally speaking, the first time that Reynolds is reported to have publically discussed any one of these water expansion proposals, he was optimistic and, perhaps a little too positive, about the potential that an idea encompassed.

Early on, Reynolds said that rain making is a "somewhat more visionary project" and that there is "good evidence" that under ideal conditions precipitation can be increased 9 to 17 percent using silver iodide generated from ground stations. He concluded that the evidence fell far short of conclusive proof and as our water administrator he said, "I am appalled by the water rights controversies that might arise from widespread application of rain-making techniques in the Southwest." Before a congressional panel in 1966, Reynolds said that he was supportive of federal funding for weather modification, but that in 1962, Dr. E.J. Workman (a scientist Steve worked for before becoming state engineer) said that seeding can actually reduce precipitation by formation of rain shadows. Reynolds also said that the practicality of these techniques had been the subject of controversy for 20 years and that he is no longer knowledgeable about the state-of-the-art of rain making. He then proceeded to prove that he knew a lot about rain making particularly by citing a then recent study that showed that cloud seeding could actually reduce rainfall.

On evaporation control using hexadecanol to form a monomolecular film on the water surface of a lake, Reynolds noted that reductions in evaporation of 23 to 73 percent were obtained in early tests. Reynolds believed that, if a water evaporation reduction of 45 percent could be maintained, the state could salvage 200,000 acre-feet per year. He found "reason for some hope" in using monomolecular film on New Mexico lakes.

Steve was a knowledgeable hydrologist and his understanding of dams and their potential for failure was even better. I was in the dam design section at the Office of the State Engineer when I worked for Reynolds in the middle 1950s and, even though I had worked on the design or construction of half a dozen dams, I came away thinking that he might just know more about the subject than I did. He took a strong interest in dam safety and found that two significant dams were unsafe, Costilla Dam and Santa Cruz Dam.

In a talk about water importation, Reynolds said, "I can't tell that the prospects for importing Mississippi water to eastern New Mexico are great, but I can say it's a proposition worth thinking about seriously and not an idle dream." There was also the possibility that the Colorado River Compact might be amended to provide more water for energy development in New Mexico. With respect to that notion, Steve said, "My legal advisor has said that pigs will fly before the compacts are amended to allow us to use California water. I am not an attorney and therefore have little understanding of the aerodynamic characteristic of pigs, but it does appear to me that we could all freeze in the dark before the necessary amendments are made."

By 1980, Reynolds began to question the feasibility of most of the water augmentation schemes that had been proposed. In 1981, when discussing weather modification, importation from other states, and desalting of brine groundwater he said, "Prudence suggests that in current planning we should not rely heavily on such augmentation schemes." In 1989, Steve concluded on the basis of physical, political, or economic reasons that neither conservation, desalination, weather modification, nor importation showed much promise as a method of increasing the available supply.

There was one technique for increasing the available supply that Reynolds never gave up on, water salvage, the reduction of water lost to non-beneficial plants. Steve would cite the canalization on the Rio Grande and low-flow channels as salvaging 80,000 acre-feet per year. He was a staunch supporter of the Bureau of Reclamation's salt cedar control project on the Pecos River. He believed that the cost of $2 million would save 26,000 acre-feet of water per year of the 600,000 acre-feet per year being lost to non-beneficial use. I worked on salt cedar control at least three times. When I worked for Steve, we even tried making paper out of those hard-wood weeds—salt cedars or tamarisk produce a high quality writing paper, but not economically. I gave up trying to cut, mow, burn, or poison them, but others continue to spend money in an effort to kill salt cedars. If Steve were alive today, he would be there cheering them on.

## Reynolds the Litigator

Steve was not only New Mexico's longest serving state engineer, he was by far our most court-oriented water administrator and he liked the label of being the "most litigious son-of-a-bitch in New Mexico" given to him by New Mexico Supreme Court Justice Carmody. Not all of Steve's battles made it to the courts. He was always ready to take issue with a person or an agency when he felt that their actions would negatively alter his ability to administer water law in the state.

Reynolds took on anybody who claimed that New Mexico didn't have enough water for this or that kind of industry. Reynolds frequently said, "It is often implied that we are at, or very near, the limit for economic development because of the scarcity of water in New Mexico, but such an implication is certainly not justified. Our economic development can be greatly extended by the orderly redistribution of water under beneficial uses under our law." Reynolds, the free-marketer, concluded that "the amount of water required for urban life is only a little more than 0.1 acre-feet depletion per year per resident. Doubling the amount of water used by industry and municipalities will only require retirement of 11 percent of water committed to irrigated agriculture." Reynolds took specific issue with representatives of federal energy agencies on this same issue. Reynolds attacked saying, "We have sufficient water to satisfy all energy development needs through 2020 as it will only take 5 to 10 percent of our available supply." What Steve was saying is all you had to do to get a sufficient water supply was to buy water rights from a larger irrigated farmer.

Reynolds took on anyone who had a different idea on how to run the Office of the State Engineer. An Albuquerque lawyer once stated that those needing to acquire water in the Middle Rio Grande Conservancy District (MRGCD) "find it impossible to deal with the irrigation district and must compete for increasingly scarce non-District Middle Rio Grande water rights." The lawyer involved then went on to suggest that the Interstate Stream Commission (ISC) take over the management of the MRGCD. The lawyer then carelessly added that "as an added benefit, granting the ISC authority over the District might well aid in avoiding on the Rio Grande the kind of costly litigation New Mexico incurred on the Pecos and the District water rights might be lost." Reynolds bristled at this and said, "The two compacts are quite different; the analogy is not well drawn" and that the water of the MRGCD was guaranteed by the Rio Grande Compact and "cannot be lost by any action or inaction" of the District Board. He then went on to enumerate the reasons the lawyer's ideas were bad. That lawyer should have known that you really shouldn't tell Steve how to do his job. Reynolds went on to say, "The District may be 'impossible to deal with,' but the Board's position in the water rights market is unfavorable at best." What

Reynolds was trying to tell that lawyer was that the price of an acre-foot of District water was too low, only $1,000 an acre-foot at that time, clearly not high enough for many farmers to sell.

Reynolds, in response to a claim by another lawyer that the state engineer has no jurisdiction over Indian water rights, said, "It is my position that the State Engineer has jurisdiction over appropriations by non-Indians on Indian lands. Without knowledge of the nature and extent of Indian rights, we must await adjudications of rights by the courts." So there is your answer!

Reynolds and I came into conflict on more than one issue after I left the Office of the State Engineer to go to the Sanitary Engineering Division in the State Health Department. Our differences arose quickly. I lobbied Reynolds for dilution water releases from Navajo Dam into the San Juan River as three towns on the river discharged treated domestic sewage above the intake for the water supply for the Navajo community at Shiprock. As he often did, Reynolds recited the sentence from the state constitution that "beneficial use shall be the basis, the measure and the limit to the right to the use of water" and then he said,"John, there is no beneficial use from dilution and you know that dilution is not the answer to pollution in New Mexico." Steve was right, but I wanted a word in edgewise so I said, "The U.S. Public Health Service is concerned about treated sewage in the San Juan reaching Shiprock." Reynolds shot back, "Well, get your sewage out of my river!" I always lost on my confrontations with Steve.

Steve was very active in the establishment of New Mexico's first set of stream standards and on the State's first water quality act. I was also deeply involved in both as head of the Governor's Air and Water Pollution Committee. While Reynolds was certainly no flaming environmentalist, he nevertheless acted on a number of water permits and proposed developments to ensure that the available fresh water was not degraded to the point that further use was prohibited. Steve was the driver to ensure that underground disposal of oil field brines did not contaminate fresh water zones. But he did believe, and inescapably so, that some degradation in water quality occurs with every use, and he did battle with, first, the Interior Department and then the U.S. Environmental Protection Agency on this issue, when federal policy on degradation of water quality became an issue.

Reynolds asked Interior Water Quality Commissioner John Quigley about words in Interior's policy that said "in no case" would degradation in surface water quality be allowed. Quigley assured Reynolds that the federal guidelines would be interpreted reasonably. Reynolds then said, "Since I was unable to find many ways to construe the words, 'in no case,' I was not greatly comforted by his assurance." Reynolds continued to argue that the words "in no case shall degradation in existing water quality occur" be taken out of the federal water pollution policy statements. Those words are still in policy statements today. To counter the federal position, Reynolds worked to get a paragraph in the State Water Quality Act adopted to the effect that "reasonable degradation of water quality resulting from beneficial use shall be allowed" and I supported Steve on the inclusion of that provision in the state act.

Steve spoke often about demands by some for maintaining a high quality environment. Once he said, "I hope that they understand that it must be paid for in higher taxes and by increased exertion of mind and muscle or they must decrease their consumption. New Mexico ranks fourth in the nation for highest percent of communities with sanitary sewer systems receiving primary and secondary treatment." I guess Steve forgot that it was the three engineers at the old Health Department who had worked hard to make that record possible.

The environmental community wanted Steve to endorse changes in the State Water Code to allow water rights for "instream flow." He was adamantly opposed saying, "There is no instream flow right in New Mexico law. You must have a man-made diversion from a stream." On the recognition of recreation and fishing as beneficial uses, Reynolds would cite an old New Mexico Supreme Court case that made both beneficial uses, and then he would unequivocally state that legislation was not needed.

Reynolds often joked about the litigious nature of our culture saying, "In our current litigious society, anything might be litigated." Yes, Steve was a litigator and once he said, "I'm sure that you know how litigious the water community has become. I have instituted a policy which requires that if somebody is not suing me, I will sue them to find out why they haven't." And you had a feeling that he might sue.

Steve kept track of his higher court cases and in speech after speech he increased his count of cases that reached the upper courts. And he would tell listeners about his winning percentage of these cases. By 1984, Steve was able to say, "I've been involved in 70 New Mexico Supreme Court cases, and we have scored about 85 percent wins in these appeals. While I won't suggest that this indicates the soundness of my decisions, one might infer that I have done reasonably well in selecting attorneys." Most of us would say, "That's a pretty good winning record, Steve." Steve's biggest win in state court was the 1960 case *Albuquerque v. Reynolds*, discussed in Chapter 6.

Reynolds was our water administrator for one-third of the history covered in the stories in this book, and lots happened during his tenure in office. Certainly, Steve was a strong-willed, forceful man, but did he look for court fights? Many of us onlookers thought so. Let's review some of the Reynolds record of going to federal court and then you decide.

Most of the federal cases involved water right adjudications where Native American water rights were involved, cases where there was some significant other federal interest, and interstate compact disputes. Here are some of the important cases in the federal courts that involved New Mexico.

- *New Mexico v. El Paso*, 1981, on El Paso's claim to the use of groundwater in New Mexico in U.S. District Court.
- *Arizona v. California*, early 1930s, on rights of various states to the use of Lower Colorado River water; it continued for many years and in 1964 New Mexico's Gila River interests were affirmed.
- *United States v. New Mexico*, 1958 federal adjudication of U.S. Forest Service claims to water rights on the Mimbres River.
- *Texas and New Mexico v. Colorado* on Colorado's under-deliveries of water required under the Rio Grande Compact 1966.
- *Texas v. New Mexico* on the Pecos on New Mexico's under-delivery to Texas, a case that started in 1974 and ended in 1988.
- *Colorado v. New Mexico*, 1978 to 1982, in which Colorado sought equitable apportionment of the water in the Vermejo River.
- *Texas v. New Mexico* on the Canadian River, on the allowable amount of storage in New Mexico reservoirs under the compact.

Other adjudication cases of interest in federal courts included the Aamodt case on water rights in the Tesuque area that started in the 1950s and ended in 2010; the Taos area adjudication that involves both Taos Pueblo rights and those of the area acequias; the Jemez River adjudication started in the 1970s, continues today and involves Native American claims; and the Upper Chama River adjudication that involved Native American rights and those on various area acequias.

Reynolds won most of these cases and he may have lost a few depending on how you keep score and who was keeping score. The Supreme Court ruled for Colorado in the Vermejo suit. My take on the case is that after appeals, New

Mexico did not suffer a loss in the Vermejo case as nothing ever happened to give Colorado any of our water. The Canadian River case was a minor and temporary loss even though Texas prevailed.

Steve's most significant victory in the U.S. Supreme Court was in the case of *United States v. New Mexico*, involving federal claims to New Mexico water by the U.S. Forest Service. Unfortunately, that case did not end the battle over federal reserved water rights in New Mexico. (see Chapter 4)

How you keep score on winning and losing court cases is not just an idle notion. *Texas v. New Mexico* on the Pecos River went to the U.S. Supreme Court. The case was bifurcated. The first trial segment was devoted solely to the meaning of the "1947 condition" in the principal apportionment provision in Article III (A) of the Pecos River Compact. In the first phase of the trial, New Mexico prevailed on the basic issue, reducing the substance of Texas' complaint from an accumulated gross liability of 1,400,000 acre-feet to 340,100 acre-feet. The consensus is that New Mexico lost but Reynolds didn't see it exactly that way when he turned on the rainbow colored lights to illuminate the best aspects of what seemed to be a dark day. Steve was not happy about Texas claims that New Mexico's water administrators exercised bad faith, and that New Mexico owed Texas $1.1 billion for shortfalls in deliveries on the Pecos River of more than a million acre-feet over the life of the compact. He was wounded by comments by Renea Hicks, Counsel for Texas, who said, "The people of the State of New Mexico can thank Steve Reynolds for having to pay $14 million." Reynolds took this as "a snide remark" and went on to say, "The case was settled for $14 million or about 1.3 percent of what Texas claimed and I think the people of New Mexico would thank the state engineer any time he can buy $1.1 billion worth of water for $14 million." In making deliveries to Texas on the Pecos River, New Mexico, Steve believed that water salvage by control of non-beneficial plants (salt cedars) would lead to compact compliance. Unfortunately, that was not to be, although Steve worked hard, with help from the Bureau of Reclamation, to make water salvage effective. After years of trying, Reynolds was personally injured by the U.S. Supreme Court's decision in favor of Texas on the Pecos River and he went on to say, "We cannot expect the Supreme Court to always be right, but it is always supreme."

Under Reynolds, the Office of the State Engineer was in both state and federal courts many times. One lawsuit that involved both state and federal courts was the El Paso case, which I was involved in while on sabbatical in 1984. I worked on case elements being developed by the Office of the State

Engineer. As Steve was going to be the decision maker, I was not allowed to talk to him about the OSE's case, and that was difficult, when we saw each other on a regular basis. I always felt that El Paso's effort to take New Mexico groundwater irritated him more than some of his other confrontations. The City of El Paso sued him as an individual, and an annoyed Reynolds said, "We are accustomed to being sued and we take it gracefully. However, I become uneasy when our neighbors get personal about these matters. I do appreciate that El Paso has not yet asked that I be denied bail." Reynolds continued, "With El Paso challenging our water laws in federal district court in an effort to take all of the water that New Mexico has below Elephant Butte, our first priority has to be to fight and scramble to keep what we have." And Steve did fight! And Steve did win most of his water wars.

# 4

# United States v. New Mexico

## Richard Simms

### Introduction

As it turned out, *United States v. New Mexico* could not have been more appropriately named. The case started as *Mimbres Valley Irrigation Co. v. Salopek*, a seemingly ordinary fight between an irrigation company and two ranchers with large state grazing leases who attempted to divert the waters of the Rio Mimbres to the detriment of downstream users. Long before the lawsuit was filed in 1966, however, the waters of the Rio Mimbres, which rise in the high, verdant mountains of the Gila National Forest, had been over appropriated by nearly 1,000 additional water users pursuant to New Mexico's doctrine of prior appropriation, and often there was not enough water to satisfy all of the water rights along the river. In this light, District Court Judge Norman Hodges saw the lawsuit as a harbinger of continued legal conflict that could be resolved only through an adjudication of all of the countervailing rights to the use of the waters of the Rio Mimbres. Accordingly, Judge Hodges ordered the state engineer to conduct a hydrographic survey of all of the rights to the beneficial use of the Rio Mimbres, ultimately including the rights of the United States, if any. Unbeknown to anyone at the time, the United States would claim a conceptually new kind of water right, a right to an *"instream flow,"* with retrospective application, that would enable the United States to preclude water from being diverted from the Rio Mimbres and applied to beneficial use on non-riparian lands under New Mexico's law of prior appropriation. The small skirmish between private parties in Deming, New Mexico, in 1966 would soon become a landmark case in the United States Supreme Court over whether the United States could usurp the plenary control it had given to each of the western states over the appropriation and use of public waters all over the West.

## The History of Federal/State Relations in Western Water Law

Until the 1970s, the United States had no water rights for any national forest, including the Gila National Forest, other than those obtained by complying with the doctrine of prior appropriation under the laws of the various western states. In New Mexico, the United States had applied for and obtained two water rights for the Gila National Forest as late as 1955. Its state-based water rights, like the water rights of anyone else, took their place in line under the doctrine of prior appropriation, giving the Forest Service a paramount right during times of shortage to those with rights junior in time. Historically, the Forest Service complied with state water law because there was no conceptual basis upon which to claim a water right under federal law until 1963. While water had been diverted from streams and rivers in the West for mining, municipal, agricultural, and industrial purposes for well over 100 years, *there was no federal water law governing its use.*

During the middle 1800s, title to most of the land in the American West had been ceded to the United States by various foreign powers, and until the later part of the century, it remained in the public domain—that is, it was unencumbered, federally owned property, subject to sale or other disposition and not "reserved" or held back for any special governmental or public purpose. There were no private rights in the federally owned land. Miners and others initially drawn to the West took up residence where they saw fit. If they needed water, they diverted and used it pursuant to local custom and law.

As the western territories and states became more populated, the United States simply acquiesced in the development of territorial and/or state water law. Turning the acquiescence into federal law in the late 1800s, Congress passed three statutes in which the United States expressly deferred to territorial and state water laws. The first was the Act of July 26, 1866, stating that appropriations of water under local custom or law would be protected. In the Act of July 9, 1870, the federal government made it clear that grantees of property from the United States would take their land subject to any and all water rights and rights-of-way for the delivery of water that had been developed under territorial or state water law. Finally, Congress passed the Desert Land Act of March 3, 1877, which, according to the United States Supreme Court, severed all of the water on the public domain from the land itself. In reviewing the history of this federal legislation in 1935, the Supreme Court held that the Acts of 1866, 1870, and 1877 caused a complete cession of the government's control over all of the non-navigable waters rising on the public domain to the western states. Accordingly, *as a matter of federal law*, property interests in land in the West were developed under federal preemption, homestead, and desert land laws, while rights to the use of water on the public domain were developed under territorial or state laws.

Recognizing that there was not enough water in the streams and rivers to satisfy the claimed rights to water in the arid West, each of the 17 western states adopted some form of the doctrine of prior appropriation, which basically ensures that each senior appropriator can fully exercise his right to divert and beneficially use public waters before junior appropriators can exercise their rights to divert and use water from the same source of supply. Given the inadequate supply, the object was to acknowledge and protect prior investments in appropriations of water from persons or entities who made their appropriations later in time. The prevailing climatological conditions in the 17 western states between the 100$^{th}$ meridian and the Pacific Ocean also dictated another unifying principle in the development of western water law—water rights could not depend solely, if at all, upon the ownership of riparian land, as they do in the states east of the 100$^{th}$ meridian. In the East, there was no need to reclaim economically unusable land by changing its nature with the delivery of irrigation water; dependable, annual rainfall made reclamation unnecessary. In the West, though, especially in the block of arid states in which riparian water rights have never been recognized (Arizona, Colorado, Idaho, Montana, Nevada, New Mexico, Utah, and Wyoming), reclamation was the only way of making most of the land economically habitable. Even in the six states along the 100$^{th}$ meridian from Texas to North Dakota and the three states bordering the Pacific Ocean, each of which has recognized riparian water rights to some degree, water rights arise not only from the ownership of land but by the act of diverting water from a stream or watercourse and applying it to beneficial use on either riparian or non-riparian lands within a reasonable time.

While tens of thousands of water rights had been developed under regimes of state water law by the turn of the century, the United States Supreme Court made a major change in the fabric of western water law in 1908 in a case called *Winters v. United States*. In 1889, Mr. Winters, one of the appellants in the case, homesteaded certain lands on the north side of the Milk River in western Montana. He and others settled on their lands, built homes, and constructed diversion, storage, and delivery works from the Milk River to irrigate their lands. In 1888, however, the United States had "reserved" approximately 30,000 acres of land for two Indian tribes. The reservation was called the Ft. Belknap Indian Reservation, located in northwestern Montana near the Idaho panhandle. The northern boundary of the Reservation was the Milk River.

The term "reserve" has a specific meaning in western water law. It describes the United States' reservation or withdrawal of specified lands from the public domain for some public purpose. With respect to the withdrawal of lands from the public domain for the Ft. Belknap Indian Reservation, it was contemplated that the Indians would restrict themselves to the area and take up farming and irrigation, but neither the Indians nor the United States had a water right to divert water to serve Ft. Belknap lands, and no mention was made of water in the treaty through which the United States created the Reservation. While Mr. Winters and the other settlers in *Winters v. United States* had perfected water rights to the use of the Milk River under Montana state law, the United States thought that there wouldn't be enough water to implement a proposed 5,000 acre Indian irrigation project on the Reservation. As a result, the United States sued Mr. Winters and the other settlers to enjoin them from maintaining their diversion works and reservoirs. Mr. Winters' defense, of course, was that he had settled in Montana pursuant to recognized federal land law and had initiated his water rights pursuant to state law, just like everyone else moving to the West.

When the case got to the United States Supreme Court in 1908, the Court paid lip service to the fact that the United States had no explicit water right for its Indian irrigation project, but the Court nevertheless articulated the Kafkaesque legal fiction that when the United States withdrew lands from the public domain in order to establish the Ft. Belknap Indian Reservation in 1888, it also must have *impliedly* withdrawn from the waters of the Milk River sufficient water needed to satisfy the purposes for which the lands were withdrawn. While no notice was ever given to the settlers on the north side of the Milk River of any competing right to use Milk River waters, the priority of the so-called "federal reserved right" was the date of the reservation, *i.e.*, 1888. Accordingly, the new federal right, created out of the blue by the Supreme Court in 1908, was paramount to Mr. Winters' 1889 rights. In other words, under the prevailing practices in the West, the new water right for the Indians would be satisfied in full in times of shortage before any water could be diverted by the competing non-Indians.

For a long time, the *Winters Doctrine* or "federal reserved water right doctrine" was restricted in legal application to Indian Reservations. A case called *Arizona v. California*, however, was filed in 1952 in the original jurisdiction of the United States Supreme Court to determine how much water each state had a legal right to use from the Colorado River and its tributaries. Subsequently, Nevada, New Mexico, and the United States were added as parties to the case. While the United States Constitution mandates that cases between or among states originate in the Supreme Court, the Court is not equipped to conduct lengthy and complicated interstate water litigation. To solve the problem, the Court has historically appointed "Special Masters" to sit as trial judges and then to report their findings to the Court. In *Arizona v. California*, the Court appointed Simon H. Rifkind, a brilliant and highly respected attorney and jurist from the firm of Paul, Weiss, Rifkind, Wharton & Garrison, LLP, headquartered in New York City, a firm specializing in corporate law, mergers and acquisitions, finance, bankruptcy and corporate reorganization, and tax law.

Judge Rifkind conducted trial from June, 1956, to August, 1958. After the testimony of 340 witnesses and the admission of thousands of exhibits, Judge Rifkind wrote a 433-page report to the Supreme Court, summarizing his findings and conclusions and recommending a decree. Because water rights for both Indian and non-Indian federal reservations were involved in *Arizona v. California*, Judge Rifkind made the somewhat rhetorical legal determination that *he could see no reason why the principle underlying the reservation of water rights for Indian Reservations was not equally applicable to other federal reservations such as National Recreation Areas and National Forests*. With no analysis or legal discussion, the Supreme Court agreed with Judge Rifkind that the *Winters Doctrine* applied equally to non-Indian reservations of land from the public domain.

With the Supreme Court's cryptic decision in *Arizona v. California* in 1963 that the United States must have implicitly reserved sufficient water to satisfy the purposes for which all non-Indian reservations were carved out of the public domain, the United States Forest Service suddenly found itself with a brand new "federal reserved right" which, in times of shortage, trumped water rights all over the West that had been perfected under regimes of state water law. While it wasn't fully understood at the time, both federal reserved water rights for Indians under the *Winters Doctrine* and federal reserved water rights for agencies of the federal government, which, according to the United States Supreme Court, are conceptually the same, have the potential for completely negating or undermining water rights perfected under state water law in two ways. The first relates to the competing priorities of the water rights, and the second relates to the ways in which Justice Department lawyers have attempted to maximize reserved rights quantitatively.

The basic principle underlying the doctrine of prior appropriation in state law in the arid West is that there will not be enough water to go around. At times, water would be plentiful, and western streams and rivers would supply most of the rights to divert and use water; at other times, however, the supply would not be sufficient to satisfy all of those who diverted the water to reclaim arid lands. Consequently, the western states created a system in which to maximize the economic benefit that would derive from the use of public waters, allowing more rights to be developed on streams and rivers than might be satisfied in times of drought, while simultaneously protecting the rights that vested earlier in time against junior rights. The doctrine of prior appropriation gives an exclusive right to the earliest appropriator to the use of water to the extent of his appropriation, without diminution in quantity or quality, whenever water is available; subsequent appropriators each have a like priority with respect to junior appropriators. It was anticipated that water rights would be regulated during times of shortage so that senior rights would be satisfied in full before any water was delivered to the junior rights.

Prior to the Supreme Court's promulgation of the *Winters Doctrine* in 1908, many streams and rivers in the western United States had been fully appropriated, and between 1850 and 1900, the doctrine of prior appropriation was adopted by statute or recognized by judicial decision in each of the 17 western states. With the Court's creation of the "reserved water right" for the Ft. Belknap Indian Reservation, however, a retroactive priority as of the date of the creation of the Reservation was imposed on the Milk River. Theoretically, the Indian water right was "reserved" only from then unappropriated water, that is, water that had not yet been appropriated under state law. Mr. Winters initiated his appropriation in 1889, but because the priority of the Indian right was retroactive, the Indian right usurped Mr. Winters' right in its entirety, along with the water rights of all of the other people who perfected water rights under Montana law between 1889 and 1908. In short, the reserved rights that were created in *Winters v. United States* in 1908 took property without compensation.

The antagonism between state-created, appropriative rights and federal reserved rights goes beyond the obvious fact that the *Winters Doctrine* takes away from what was thought to have been relegated to the plenary control of the states. With the Court's extension of the doctrine to non-Indian reservations in 1963, however, the antagonism was magnified because appropriators under state law had no notice until 1963—even by legal fiction—of competing federal claims to the same water they had appropriated. Before then, they had sound legal reason to believe that water was available to make their appropriations, and they could not have reasonably expected that a paramount interest in the same water might be claimed in the future. Compounding the problem, the reservation doctrine provides enough water to satisfy *future* as well as *present* water requirements on a reservation, thus, in a fully appropriated system like the Rio Mimbres, permitting the United States to make a new appropriation long after 1963 with the priority of the original withdrawal from the public domain, effectively taking without compensation all water rights predicated upon intervening uses.

The other major problem in federal/state water law jurisprudence in the West derives from excessive quantitative claims for federal reserved rights, especially the claims made by Justice Department lawyers for Indian reservations. When Judge Rifkind addressed the quantity of water reserved for five Colorado River Indian reservations in *Arizona v. California* in 1963, he reasoned that the reserved rights should not be formulated to satisfy the number of Indians residing on the reservations when the reservations were created. He also reasoned that the reserved rights doctrine could not be blind to the fact that it might be necessary to expand the agricultural and related needs of the reservations. Accordingly, Judge Rifkind utilized a standard that would facilitate the quantification of the rights, notwithstanding potential population growth, while simultaneously establishing water rights of a fixed magnitude and priority in order to provide certainty for all water users. He concluded that when the United States reserved the lands from the public domain, it impliedly reserved enough water to irrigate *all of the practicably irrigable acreage* on the Reservations. Under what became known as the "PIA" standard, the "proof" of a *Winters* right consisted of a benefit-cost analysis of one or more proposed irrigation projects in which the annual benefits exceeded the costs. In *Arizona v. California* itself, the standard resulted in an award of approximately 906,000 acre-feet annually to five reservations having a combined population of 2,940 Indians. The remaining 12,600,000 acre-feet of the average annual flow of the Colorado River water were apportioned to the 23,658,000 people then living in the Colorado River Basin states of Arizona, California, Colorado, Nevada, New Mexico, Utah, and Wyoming. As a result, on a *per capita* basis, the use of practicably irrigable acreage as the standard of quantification was 579 times more favorable to the Indians than the non-Indians. Because the population on the Indian reservations has since declined, and the population in the Colorado River Basin states has increased dramatically, the unwarranted award to the Indians has become all the more lopsided.

In the 1970s and 1980s, the quantitative extent of Indian water rights was being hotly litigated in the West. During this period, the United States added a social and ethical bias to its benefit-cost analyses in order to compensate for perceived, past injustices to Indians. With the increase in Indian water rights litigation, the benefit-cost analyses employed by the United States became charades to generate excessive and unnecessary awards of water. At the same time, the Justice Department lawyers began to assert that the purpose of every Indian reservation was to create "a permanent homeland," a concept designed to overcome the limitation to a need to satisfy the actual, historical purposes for creating a given reservation. Such a concept is of no utility, however, in explicating the implicit need for water. Unless the purpose of a reservation is specific enough to determine by rational inference a certain need for water, the *Winters Doctrine* becomes nothing more than a black hole in the $5^{th}$ Amendment through which water rights perfected under regimes of state law can be taken without compensation. The last case litigated in New Mexico, a case called *State ex rel. Martinez v. Lewis* that was finally decided in 1994, is a good example. In that case, the United States claimed almost 18,000 acre-feet annually out of the Rio Ruidoso stream system for a proposed irrigation project for the Mescalero Apache Tribe when the average annual flow of the Rio Ruidoso was 9,600 acre-feet. Below the Reservation boundary, the Ruidoso and its tributaries had been fully appropriated by a largely Spanish-American community well before the Mescalero Indian Reservation was reserved from the public domain. The benefit-cost analysis that was done for the proposed irrigation project was a farce, and everyone involved in the case knew that neither the United States nor the Indians intended to build the project, but instead intended to "lease" the water rights that they hoped to get to the downstream users. The entire case was put together simply to generate money. As Wendell Chino, then the Chairman of the Mescalero Apache Tribe, put it, 'it's about time that [the non-Indians] paid for our water.' Fortunately, the Justice Department lost the case in a decision written by LaFel Oman, a retired Chief Justice of the New Mexico Supreme Court.

Historically, the development of western water law made perfect sense, at least initially. Through the Acts of 1866, 1870, and the Desert Land Act of 1877, the United States divested itself of control over all non-navigable waters on the public domain in the West by relinquishing plenary control to the states and territories. In 1908, though, the United States Supreme Court promulgated a fundamental change in what Congress had done by *creating* federal reserved water rights for Indian reservations, federal water rights that usurped water rights perfected under state law pursuant to the scheme that Congress had put in place. In 1963, the Supreme Court did the same thing, exacerbating the mess it had earlier created by extending the reserved water rights doctrine to all withdrawals of land from the public domain for any federal purpose, giving numerous federal agencies new water rights that would further upend the federal government's historical deference to state law. It was out of this rather aimless historical development that *United States v. New Mexico* emerged.

## Luna County District Court to the United States Supreme Court

When the State of New Mexico, in the form of Steve Reynolds, the New Mexico State Engineer, intervened in *Mimbres Valley Irrigation Co. v. Salopek* in 1971, the case grew from three parties to 1,000 parties in a statutory stream system adjudication in which the water rights of all the water users in the stream system would be adjudicated, including the claimed rights of the United States. The State of New Mexico's rendition of the United States' water rights was first submitted to the district court of Luna County in the hydrographic survey that Judge Hodges ordered the State Engineer to prepare. At the time, the United States was involved against individual western states in two kinds of water litigation, *i.e.*, substantive litigation over the nature and extent of the United States' reserved water rights, and procedural litigation over whether the federal reserved rights of the United States should be adjudicated in federal district court or in state district court under the McCarran Amendment, a federal statute that waived the United States' sovereign immunity with respect to stream system adjudications in the West. The procedural issue was never litigated in regard to the Mimbres adjudication itself, but New Mexico was a leader in the West in winning lawsuits upholding the right of the individual states to join the United States as a party to adjudications in state district court. Steve Reynolds, once dubbed the most litigious SOB in New Mexico's history, won lawsuit after lawsuit either keeping the United States in state district court adjudications or throwing it out of adjudications it had filed in federal district court. In *Mimbres Valley Irrigation Co. v. Salopek*, the fight went straight to the substance of the United States' federal reserved water rights.

While there were nearly 1,000 parties in the expanded adjudication, the water rights of each party in an adjudication are first determined either by stipulation or litigation between the State of New Mexico and each individual water user. Subsequently, the rights of individual parties may be challenged by other individual parties. Because the adjudication suits are so large, often involving thousands of defendants, Special Masters are usually appointed to handle the initial disputes between the State of New Mexico and each defendant. In the *Mimbres* case, Irwin S. Moise, a retired Chief Justice of the New Mexico Supreme Court who had authored many of the Court's water law opinions, was appointed as Special Master. The first issue that he addressed arose out of the initial pleading in which the United States set forth the legal underpinnings of its claimed water rights. The Forest Service at the time had not yet developed any firm guidelines regarding the way in which it would claim its new "federal reserved rights" to the use of water in any of the adjudications in the West. In its Answer in *Mimbres Valley Irrigation Co. v. Salopek*, the United States described its reserved rights deriving from the various reservations of land between 1899 and 1910 that comprised the Gila National Forest in the general dogma of the *Winters Doctrine*, namely that when the United States reserved the lands from the public domain for the Gila National Forest it impliedly reserved sufficient waters of the Rio Mimbres to satisfy the purposes for which the lands were withdrawn. At the initial pretrial conference in 1972, it was learned that the United States needed additional time to identify and describe its claims to present water rights in the Gila National Forest, consisting mostly of alleged rights for domestic uses and stock ponds, and to set forth in detail its claimed rights for future uses. Based upon the agreement of parties, Justice Moise afforded the United States additional time within which to describe its water rights claims with more specificity.

By 1972, when I was hired by Paul Bloom as an attorney at the Office of the State Engineer after clerking for Chief Judge Oliver Seth of the Tenth Circuit Court of Appeals, there had been essentially no development in the law of federal reserved water rights. Two cases from Colorado had been decided by the United States Supreme Court. In a case called *United States v. District Court in and for Eagle County*, a McCarran Amendment case in which the United States asserted that it could be made a party to a state court adjudication only in regard to water rights it had acquired under state law, Justice William O. Douglas wrote the Court's opinion, holding that the United States was amenable to state court jurisdiction with respect to all of its water rights, including its federal reserved water rights, if any, for the White River National Forest. In passing, Justice Douglas noted that the United States had often reserved water rights based on withdrawals from the public domain and that the reservation of waters could only be implied in an amount needed to satisfy the purposes of the reservation of land. Justice Douglas also wrote the Court's opinion in *United States v. District Court in and for Water Div. No. 5*, a much broader companion case involving the United States' claims in the streams and rivers in Colorado that were tributary to the Colorado River. The case involved the federal reserved water rights claims for four National Forests, the Department of the Interior, the Bureau of Reclamation, the National Park Service, the Bureau of Sport Fisheries and Wildlife, and the Department of the Navy. The Court denied the United States' attempt to evade state court jurisdiction, and nothing was said about the substance of federal reserved rights beyond Justice John Marshall Harlan II's statement in his concurring opinion that he didn't think that the United States had reserved rights, either in general or with respect to all of the claims involved in the case. A third case had been filed by the United States in federal district court in Nevada to enjoin an extended ranching family named Cappaert from pumping groundwater that allegedly lowered the water level in an underground pool of water containing a rare species of fish in Devil's Hole, a remnant of the prehistoric Death Valley Lake System. In *United States v. Cappaert*, which wasn't decided and appealed until 1974, the United States claimed reserved rights that allegedly arose under the American Antiquities Preservation Act.

*Mimbres Valley Irrigation Co. v. Salopek* was the first case of any substance that I undertook after I went to work for Steve Reynolds in 1972. As usual in reserved rights cases, the United States made the largest quantitative claims it could, initially asserting that the needs of the Gila National Forest had expanded greatly since 1899 when the Forest was first reserved from the public domain and that the Forest Service's water rights should be quantified based on the current and future needs of the Forest as perceived by those in charge of its administration. I had read all of the cases relating to the creation and development of the *Winters Doctrine*, and I strongly disagreed with the position taken by the United States. In my first brief in the case, which had been reviewed once by Paul Bloom, but not by Steve Reynolds, I took the position that reserved rights for national forests could arise only for the purposes for which the United States could reserve forest lands from the public domain under the Organic Administration Act of 1897, the only federal legislation that authorized the President to withdraw lands for national forest purposes. Not knowing what I was talking about beyond the fact that I had spent a good deal of time enjoying the recreational benefits of various national forests, I conceded toward the end of the brief that recreation was a legitimate use of forest lands. As fate would have it, Steve Reynolds reviewed the brief after it was filed. Reynolds reviewed everything he read with a pencil in his hand, and he returned his copy of the brief to me, stating that he thought the brief was well written, with one exception. In the upper right-hand corner of the first page of the brief, Steve had penciled, "a use is not a purpose." My heart sank.

It took the United States more than a year to catalog its alleged, present uses of Rio Mimbres waters and to prepare its claims for future uses. During trial of the case in 1973, the United States produced its list of claimed rights for forest uses, including recreational, wildlife, and aesthetic uses. The United States also claimed three "instream flow rights" for "fish" purposes, which it later claimed served the additional purposes of erosion control, fire protection, watershed protection, and wildlife habitat protection.

Instream flow water rights were unheard of in the western states until about 1970. The concept was anathema to the doctrine of prior appropriation, which required an intent to appropriate, a diversion of water, and application to beneficial use with reasonable diligence. Historically, the diversion requirement was thought to prevent speculation in the public water supply by compelling actual expenditures. It was also a part of the theoretical basis of the doctrine of prior appropriation, deriving from the fact that the ownership of property had nothing to do with perfecting a right to divert water and convey it across the property of others for application on non-riparian lands. An appropriative right is usufructuary in nature or a right to use public waters, that is, a right that can be exercised by diversion and use. An "instream flow right," though, cannot be exercised. *It is simply a name or a misnomer for the power to preclude: 1) the diversion and use of public waters in the first place, 2) a transfer of the place of use of an existing right from downstream of the instream flow to upstream lands, or 3) the exercise of an appropriative right that had been transferred upstream before the instream flow was recognized.* In the context of *Mimbres Valley Irrigation Co. v. Salopek*, the only utility of the United States' claimed instream flow rights would have been to prevent any new appropriation of water under New Mexico law on private lands situated upstream in the Gila National Forest, prevent the transfer of appropriative rights to lands upstream of the instream flow, or prevent the exercise of an appropriative right that had been transferred to lands upstream. Indeed, the latter is what the Forest Service intended to do according to a letter from the Regional Forester to Steve Reynolds. Phelps Dodge Corporation, the largest employer in Grant County, had invested some $425,000,000 in a large copper processing mill at Tryone, New Mexico, on lands above a portion of the Gila National Forest. To operate the mill, Phelps Dodge had purchased and transferred under New Mexico water law some 11,000 acre-feet of downstream water rights, with priorities ranging from 1882 to 1961, to the processing mill and to a point of diversion above a claimed instream flow. Without mincing words, the Regional Forester told Reynolds that Phelps Dodge's rights would be shut down if their exercise interfered with instream flow requirements necessary for fisheries. While the United States is fond of claiming that instream flows are "non-consumptive," they can be 100 percent confiscatory. With the United States' claim for instream flows in *Mimbres Valley Irrigation Co. v. Salopek*, it was obvious that we had a major problem on our hands. It also would have been a major problem for each of the other western states if such a water right were recognized by the United States Supreme Court.

Under the Organic Administration Act of 1897, the only purposes for which a national forest could be reserved were to improve and protect the forest lands in order to secure favorable conditions of water flow for appropriators, *i.e.*, to manage the watershed to maximize the water available for appropriation under state law, and to provide a continuous supply of timber products. Forest lands could be used for other purposes, but they could not be withdrawn from the public domain for other purposes. When Reynolds' cryptic note that a use is not a purpose sank in, it became clear to me that my concession that the Gila National Forest could be used for recreational purposes could open a whole box of Pandoras, as Governor King would say. In the proceedings before Special Master Moise, I wasn't able to completely retract the concession, however, but only to limit it to recreational uses incidental to hiking, fishing, camping and hunting, a concept, in my opinion, that the United States still could easily exploit. Justice Moise did agree with New Mexico on the fundamental legal principles that the Gila National Forest could have been reserved only to manage the watershed in order to maximize the water supply for use under state law and to secure a continuous supply of timber. While he also agreed that instream flow rights could only be used in derogation of water rights perfected under the doctrine of prior appropriation, he recommended in his report to Judge Hodges that the instream flow claims should be recognized as long as they did nothing more than prevent the transfer of appropriative rights to upstream, private lands within the Forest. He also recommended that the stockwatering rights should be awarded to the Forest Service as opposed to the permittees who were actually watering stock.

The only way to "appeal" the recommendations of a Special Master, either to a district court judge or to the United States Supreme Court in original actions, is to "take exception" to the Special Master's Report. I was in an awkward position in taking exception to Justice Moise's report to Judge Hodges given the fact that Justice Moise, as a former Justice of the New Mexico Supreme Court, had previously been the ultimate arbiter of whether some of Judge Hodges' decisions were correct. In *Mimbres Valley Irrigation Co. v. Salopek*, the tables were turned. In our exceptions to Justice Moise's report, we contended that the United States was not entitled to reserved rights for minimum instream flows, for recreational uses of any kind, or for any kind of water use made by private users as permittees of the Forest Service.

The case came before Judge Hodges on New Mexico's objections to Justice Moise's Report in 1976. After briefs and extended oral argument, Judge Hodges, a man with both feet on the ground, who in my opinion was the real hero in the case, issued his decision upholding New Mexico's objections to Justice Moise's Report. He concluded that stockwatering rights should be awarded to the permittees instead of the Forest Service, that recreation was not a purpose for which reserved rights could be recognized, and that the Forest Service had no rights to instream flows for fish or any other purpose. In my view, the integrity of the law was upheld in all respects, including the elimination of my mindless concession that recreation was a legitimate use of national forest lands.

On July 2, 1976, the United States appealed *Mimbres Valley Irrigation Co. v. Salopek* to the New Mexico Supreme Court. During the progression of the case from Special Master Moise through Judge Hodges, there had been only one significant reserved rights case in the United States Supreme Court, *Cappaert v. United States*, the case in which the United States claimed a federal reserved right to preclude the Cappaerts from drawing down the level of the pool in Devil's Hole by pumping hydrologically related groundwater. The Cappaerts had lost in federal district court in Nevada and lost again on appeal to the Ninth Circuit Court of Appeals. The Supreme Court granted *certiorari* to consider the scope of the implied reservation of water rights doctrine. In *Cappaert*, the Supreme Court upheld the United States' reserved right, but explained that a reserved right can arise only in the amount of water necessary to fulfill the purpose of the reservation, no more. Accordingly, the Court concluded that the district court had tailored its injunction to minimal need, curtailing the Cappaerts pumping only to the extent needed to preserve an adequate water level in Devil's Hole. In our brief in *Mimbres*, the *Cappaert* decision was added in support of New Mexico's position that the Gila National Forest could only have been reserved to manage the watershed in order to maximize the water available for appropriation under state law and to provide a continuous supply of timber. Our position, simply, was that only the two purposes of the reservation could form the bases of the Supreme Court's explication of whether water was needed. Uses of forest lands played no role in that analysis.

The panel of justices on the New Mexico Supreme Court consisted of Justice H. Vern Payne, Chief Justice LaFel Oman, and Justice Dan Sosa, the three appellate judges in New Mexico for whom I developed the most respect in some 40 years of practicing water law—not because of their decision in *Mimbres*, but because of the way in which each of them had handled numerous other water law cases. In *Mimbres*, Justice Payne wrote the Court's opinion, upholding Steve Reynolds' understanding of the law of federal reserved water rights for National Forests on all counts.

On October 3, 1977, the United States petitioned the United States Supreme Court for a Writ of Certiorari, the procedural means through which the United States could appeal the decision of the New Mexico Supreme Court. Because the United States was the petitioner in the United States Supreme Court, *Mimbres Valley Irrigation Co. v. Salopek* suddenly became *United States v. New Mexico*. When we filed our brief for the State of New Mexico, all of the other states west of the 100$^{th}$ meridian joined in a single *amici curiae* brief in support of New Mexico. At this juncture, the name of the case mirrored what the case was ultimately all about, *i.e.*, the extent to which the Justice Department lawyers could rewrite history in order to reclaim the plenary control over non-navigable western waters that Congress had given to all of the western states.

Once the Supreme Court granted *certiorari*, I had to begin the preparation of the most important brief I had ever written. Until then, I had done all of the research and writing in the case by myself, and in the end I had apparently done reasonably well, given the status of the case. But something was missing. The case would turn on two things in my mind, namely 1) the legislative history that demonstrated that the Gila National Forest could have been reserved only to ensure favorable conditions of water flow to appropriators under state law and to provide a continuous supply of timber, and 2) a conceptual understanding that the recognition of additional federal reserved water rights had the effect of taking water rights perfected under state water law without compensation. With respect to the legislative history, I had done considerable research in regard to the Organic Administration Act of 1897, but not enough. I had also done no research into other forest-related legislation or other federal legislation under which federal reserved rights might be claimed that might contrast with and shed additional light on reserved water rights for forests. Consequently, I assigned the job of investigating all of the legislative history to Don Klein, a very smart lawyer, educated initially at the University of Chicago, whom I had just hired. I also hired another, extremely bright lawyer who would ultimately become my partner in private practice, Jay Stein. Jay's undergraduate degree was in history. Mine was in English. The three of us had less than three months, *i.e.*, until March 31, 1978, to do all of the needed research and write New Mexico's brief, a printed "brief" that turned out to be three-quarters of an inch thick.

Klein tore into the work like a madman in the open stacks of the New Mexico Supreme Court library. He couldn't have selected a better place in which to do the research because the Supreme Court Library was one of few federal repositories in the country containing all of the documents to which reference is made in the Congressional Record. He may as well have been in the National Archives. Working sporadic, endless hours, Klein roamed the stacks in the library and summarized his research in countless memoranda, memoranda that were so numerous and complex that I had difficulty keeping track of them. We were also pressed for time. In this regard, Jay Stein was given the responsibility of organizing the memoranda, reviewing them, and piecing together a condensed, lucid history of the two purposes for which forest lands could be reserved from the public domain, a history that would become one of two main points in New Mexico's brief before the Supreme Court. In the end, we cited and explained the relevant provisions of 36 federal statutes, 42 passages from the Congressional Record, six House Executive Documents, 18 passages from House Reports, five contemporaneous Department of the Interior Decisions, and passages from 13 Senate Documents, references to all of which could later be found in Justice William Rehnquist's decision in *United States v. New Mexico*. This time, *the United States Supreme Court* upheld Steve Reynolds' understanding of the law of federal reserved water rights, limiting the Forest Service's rights to the minimum needed to satisfy the two purposes for which lands could be withdrawn to create forests, holding that the stockwatering rights should be adjudicated to the permittees who grazed stock, as opposed to the Forest Service, and denying the United States' claim to instream flows for "fish" purposes.

The United States Supreme Court's "term" or the part of the year during which it hears and decides cases always begins on the first Monday of October and almost always ends on the last day of June. I had worked very hard on

*United States v. New Mexico* and scores of other cases during my first six years of working for Steve Reynolds, and after the catharsis that came with my departure from the Watergate Hotel, where I had spent a week preparing for oral argument in *United States v. New Mexico*, it was about time to take a vacation. Along with four other attorneys, an architect, and our spouses or girl friends, I had planned an extended trip around the Sea of Cortez in a 50-foot power boat, an old, somewhat dilapidated wooden boat with a two-way radio that worked only at dawn on alternate Tuesdays. We boarded the boat on time on July 2, 1978, by way of a dingy that picked us up, two at a time, on a beach in San Carlos, Mexico. The United States Supreme Court, however, had not been on time, and it did not release its opinion in *United States v. New Mexico* until July 3rd.

After returning to San Carlos on July 13th and driving up to Hermosillo, I finally managed to call Steve Reynolds from a phone booth outside of the motel where we were spending the night. Steve, a five-unit a day diabetic who sometimes slipped into insulin shock, answered the phone in insulin shock. All he could say after I asked him who won the case was, "Richard, you won the *whole* enchilada . . ." With that, he hung up. When I returned to work two days later, I first saw Steve in the hallway of the Bataan Memorial Building near the entrance to the Office of the State Engineer. He was not in insulin shock, but he was nearly as cryptic as he was when I called him from Mexico: "Well, Richard, we're back in business."

## Epilogue

With the United States Supreme Court's decision in *United States v. New Mexico* on July 3, 1978, New Mexico won a hard-fought, protracted battle over the control of non-navigable waters rising on the public domain in the West. New Mexico, however, did not win the war, either for itself or the other western states. In one way or another, the United States is further diminishing state control over western waters through the passage of environmental, wildlife and wetlands protection, and water quality legislation. For example, the extent of the federal government's power over navigable waters under the Commerce Clause had historically depended on the definition of navigable waters. In a Supreme Court case decided in 1825, it was held that the English common law test of whether the water in question was affected by the ebb and flow of the tide answered the question—all inland waters above the influence of the tide were non-navigable. Today, as one commentator put it, the test of navigability under the Clean Water Act is whether the body of water, no matter where or how small, will float a Supreme Court opinion.

The law today is much different than it was a century ago. The Endangered Species Act, the Clean Water Act, the Safe Drinking Water Act, and numerous other pieces of federal legislation all have the effect of creating "regulatory water rights" or federal prerogatives over non-navigable water that either reform, destroy, or diminish water rights perfected under regimes of state water law. Compounding the problem is a growing disinterest or lack of education on the part of the Attorneys General and Assistant Attorneys General charged with the responsibility of protecting state interests. Some don't even know what *United States v. New Mexico* was about, much less the legal history that gave it life. Many are members of the Western States Water Council, an assembly of states' attorneys initially created in the 1960s to share thoughts and ideas in ongoing water litigation against the federal government. Today, it is a politically correct organization wherein all of the antagonistic parties in the continuing legal confrontation over western waters gather at quarterly meetings to pat one another on the back. In short, many attorneys responsible for protecting the interests of the western states have either given up or don't know better.

In this regard, when the Reclamation Act was passed in 1902, § 8 stated that nothing in the Act shall be construed to affect or to in any way interfere with the laws of any state or territory relating to the control, appropriation, use, or distribution of water used in irrigation. The same is true of most modern-day water quality, wildlife protection, and environmental legislation. Each Act contains similar language. Once the language is put into the legislation, however, the states seem to go to sleep at the wheel. The Justice Department lawyers, however, remain as eager as ever to superimpose their modern-day environmental and ecological ideas on a legal history that didn't recognize them.

# 5

# "A Dog of a Lawsuit"
# Texas v. New Mexico

### Em Hall

I

The dispute between Texas and New Mexico over the shared, scarce, and unpredictable water of the Pecos River was full of anger at the start of the 20th century and gained nothing but rancor for almost the next hundred years. By the time it turned into a 14-year-old U.S. Supreme Court lawsuit sputtering on between 1974 and 1988, the dispute was a mad maelstrom of scientific uncertainty, legal posturing, bitter personality conflict, and high-stakes gambling. When in a private memo to his fellow justices, Supreme Court Justice Byron "Whizzer" White called *Texas v. New Mexico* "a dog of a lawsuit", he didn't know how right he was.

Who could make sense of a nasty, brutish, and long fight like that?

I tried as a young lawyer working for State Engineer Steve Reynolds in the early stages of the litigation and couldn't make heads or tails of the emerging struggle. I tried again after it was all over when I wrote an entire 100,000-word account of the battle in a book UNM Press published in 2002 under the name "*High and Dry: The Texas-New Mexico Struggle for the Pecos River.*" *High and Dry* told readers everything they needed to know about the lawsuit and, some critics said, a lot more. Despite its length and detail, the book ended in a final chapter that simply expressed the perplexity that I'd always felt about the lawsuit and the River. Some readers praised the last chapter as the equal of a *New Yorker* essay. Others wondered what the hell that chapter was doing there, anyway.

Now, almost ten years later, the Pecos River finally has quieted down enough for me to see more clearly what happened in *Texas v. New Mexico*. The lawsuit did not so much clearly define the rights of two bitter state enemies over a shared river as it did create a whole new regime for the River that seems to have worked. This uneasy new regime may or may not endure, but it does suggest that the role of the lawsuit was not so much to make the law of the River clear as to prod upstream New Mexico to adjust its uses and contributions in such a way as to meet rigid established interstate obligations and also to stretch and bend its own internal laws. In a world like that, it's silly to wonder who won the lawsuit. Everybody won and everybody lost and the world was changed in the process. It's the change, not the victor, that counts.

II

At the beginning and in the end, *Texas v. New Mexico* was about the 1948 Pecos River Compact, that interstate contract by which two states apportion the waters of a shared river between them. Compacts like this make the fundamental law of the river.

On the one hand, upstream states, like New Mexico on the Pecos, usually favor depletion-based compacts. These compacts specify how much water an upstream state can consume and give the rest, whatever it may be, to the downstream brothers. Depletion-based compacts shift the vagaries of rivers and the imprecision of their scientific descriptions downstream to the states below.

On the other hand, downstream states like Texas favor flow-based compacts, ones that obligate an upstream state to deliver a certain amount of water passing an upstream gage to a downstream gage near the state line. Flow-based compacts make the upstream states responsible for everything that happens in between the two gages, including unpredictable nature and approximate science.

Ten years before the Pecos River Compact, Texas and New Mexico (along with Colorado) had entered into the 1938 Rio Grande Compact. That compact was a classic flow-based compact obligating New Mexico to deliver to Texas a set portion of the Rio Grande flowing past an upstream gage on the River north of Santa Fe. Ten years later, in the 1948 Pecos River Compact the same two states started out making a depletion-based compact. "New Mexico shall not deplete by man's activities . . ." the critical apportionment provision of the Compact opened, obligating New Mexico not to deliver a set amount of water to Texas but to rein in human uses of the river so the rest, whatever it was, could make it to Texas.

But to that fundamental depletion-based obligation the 1948 Pecos River Compact added a flow-based baseline, "the 1947 condition of the River." This was no Rio Grande Compact, two gage measurement of Pecos River flows. Instead it was the equivalent of a 48-gage measurement, with more than 100 points of inflow and outflow, adjusted for the variations in annual inflows and outflows all along the Pecos River. If in any year New Mexico delivered to the state line less water than the "1947 condition" indicated there should have been, then in that year there was a "departure" from the "1947 condition." And if there were a departure, then New Mexico would be responsible for that part of the departure that was attributable to man's activities in New Mexico.

Got it?

It took more than 14 years of Supreme Court litigation to get it. However, Royce Tipton, the Denver-based water engineer who came up with the hybrid depletion/flow apportionment provision in the original Pecos River Compact, justified it another way. The provision operated the way that nature operates, the silver-tongued Tipton said over and over again. In the course of the litigation, law transformed nature until there was nothing left but a fairly mechanical legal obligation.

## III

In 1974 West Texas lawyer and entrepreneur R.B. Magowan joined the Pecos River Compact Commission and the interstate Pecos world was never the same again. On joining the commission, Magowan found himself working in a forgotten world of Austin-based Texas politics, but that didn't stop him from raising louder and louder hell about the diminishing amount of Pecos River that was reaching West Texas Pecos River farmers. It had all been settled, said Magowan, by the 1948 Pecos River Compact.

The Compact guaranteed that New Mexico would not further deplete the Pecos River beyond the amount that it was depleting it by man's activities in 1947. Magowan knew that feverish New Mexico well drilling just before and well after the Compact went into effect, especially in the area just north of Roswell, probably caused the kind of increased depletions that the Compact forbad. He even guessed that the delayed effect of some wells existing before 1947 might reach the River after 1947 and therefore might also be prohibited. But when he looked into it, R.B. Magowan found an even bigger bomb: Using the Compact's own original definition of the baseline 1947 condition, it was relatively easy to show that over the first three decades of the Compact New Mexico had under-delivered Pecos River water to the Texans by almost 1,000,000 acre-feet of water. That was a lot more water than the Pecos River produced in a year and just the kind of hard, larger-than-life number that the larger-than-life Texas commissioner could point to. To a trial lawyer like Magowan, here in this number was a lawsuit just dying to happen.

Magowan's bluster masked a couple of fundamental problems with the claim. One was that such a suit would chart unknown seas as the first suit in the United States to ask for damages for the violation of an interstate water compact.

Who knew if it could even be done?

In addition, the Pecos River Commission established by the Compact itself had adopted in 1963 a new measure of the 1947 condition baseline that more accurately reflected conditions on the River in that year. The new definition significantly reduced the under-deliveries by New Mexico to Texas.

Who knew if the Commission could have done it?

To that altered definition of the baseline 1947 condition, the 1948 Compact only made New Mexico responsible to Texas for increased depletions by man's activities.

Did that mean that the Compact guaranteed Texas no water and only restraint on the part of the upstream state?

Finally, the Magowan Texas-style bluster masked the fact that just as the interstate war between Texas and New Mexico surfaced an equally contentious battle arose between the intrastate Carlsbad Irrigation District (CID) and Roswell area Pecos Valley Artesian Conservancy District (PVACD). That battle had simmered for even longer than the battle between Texas and New Mexico. Except for the minimal and marginal rights of the ancient Hispanic communities near the Pecos River headwaters, the CID had the oldest surface water rights on the lower Pecos River, established at the turn of the 20th century and superior in time to the groundwater rights later established in the prosperous Roswell artesian basin. For the better part of the 20th century, CID had complained bitterly that the PVACD wells had sucked the base flow of the nearby River dry before it got to CID where it belonged. For nearly as long, the New Mexico state engineers had worked hard to avert a CID priority call on the River. Finally in 1976, just as *Texas v. New Mexico* was getting underway State Engineer Steve Reynolds could hold the CID at bay no longer. The district formally called the priorities on the Pecos River, asking Reynolds to shut down the upstream PVACD wells so the water could reach Carlsbad.

The 1976 CID priority call was the first ever on the Pecos River and indeed the first on any major New Mexico stream system. To complicate matters, the CID call was the first in the West to ask a court to shut down junior, upstream groundwater diverters to get wet water to senior, downstream surface water rights owners. Some western states, Texas included, treated groundwater and surface water as completely separate resources. Others assumed that, even if they weren't legally separate, you still couldn't shut off a junior groundwater right to supply a senior surface water right because, even in a stream-connected aquifer, shutting off a junior well wouldn't produce water in the river quickly enough to make priority enforcement viable.

What was State Engineer Steve Reynolds to do about this local Pecos River problem at just the same time that the Texas-New Mexico interstate squabble was really heating up?

In typical ground-breaking manner, Reynolds came up with a solution in 1975–1976. He would drill state wells near the River and near the most junior of the Roswell wells. If the CID was short, he would pump those state wells, put the water in the River and send it downstream to the CID storage dams. Then in the next year, he would shut off the junior Roswell wells in the amount that he had had to pump the state wells to offset the impact of that pumping. The senior CID would have its water. The junior Roswell wells would have ultimately paid for it. And the state wells would have overcome the timing lag in supplying senior surface water rights with junior groundwater rights. It was a brilliant engineering solution, true to New Mexico's explicit Constitutional guarantee that senior water rights would be superior to junior water rights.

In the end, however, Reynolds couldn't bring himself or New Mexico to do it in 1976. It was politically too complex within southeastern New Mexico. Texas was after him full bore in the United States Supreme Court. So, in the face of that uncertainty, he bought time, a lot of time.

For technical legal reasons, Reynolds decided that he couldn't enforce any water rights on the Pecos River until they all had been formally adjudicated, a task that has not been completed today, more than 35 years later. With the CID priority call on the backburner, Reynolds turned his full attention to the Texas lawsuit asking for Pecos River Compact enforcement in the United States Supreme Court. In that forum Reynolds for the first decade of the lawsuit adopted the same solution that he ultimately had come up with by way of the CID priority call: He bought time.

## IV

Between 1974 and 1983, Steve Reynolds and his lawyers did everything they could to impede and slow down the Supreme Court suit. As the upstream state threatened by the claims of a downstream state, New Mexico was served by delay just fine. Reynolds' lawyers stalled, fought the Texans on technical grounds, crowded the proceedings with engineering data galore, and made of the lawsuit a nightmare worthy of Charles Dickens at his most cynical.

The Supreme Court had appointed a "special master" to hear the case for them and had chosen an experienced judge and water lawyer, Denver-based Jean Breitenstein, who actually had written the Pecos River Compact in 1948. Even for him, the early proceedings left him confused, exhausted, and frustrated. "The engineers, God bless them," he quipped early on, "I love them, I have represented them, but by means of the formulas that they use, they can take the same basic data and come out one to the North Pole, the other to the South."

After almost ten years, by 1984, Breitenstein had gotten only a couple of basic issues straightened out and the supervising Supreme Court had approved. He'd tossed out Texas's basic theory that the baseline "1947 condition" was cast in cement by the Compact itself. So much for R.B. Magowan's original Texas theory that had yielded that 1,000,000 acre-foot New Mexico debt to Texas. But Breitenstein also had ruled that the Compact commission's 1963 review of basic data wasn't binding either. The "1947 condition" of the Pecos River remained to be figured out. Texas and New Mexico could now do it themselves or, if they couldn't agree, God forbid, then the Court would do it for them. If the previous decade had been any indication of the speed of progress to be expected, this might take forever. The exhausted 78-year-old Breitenstein quit.

Fearing that the Court or its master might do worse, Texas and New Mexico actually managed to agree on an acceptable definition of the "1947 condition" on the Pecos River. None of the engineers who worked on the new values to be assigned to the 48 inflow/outflow values that would account for River flows had complete confidence that the new "1947 condition" did more than come a little closer to the "way nature actually operated" on the River, but the new agreed-on values were better. The new "1947 condition" description seemed to apportion the flood flows of the River between the two states 50/50.

And guess what? The new values reduced the amount of water that New Mexico had under-delivered to Texas in the 1948-1983 period from the flabbergasting 1,000,000 acre-feet that Robert Magowan had first claimed to just over 300,000 acre-feet or about 10,000 acre-feet a year. New Mexicans didn't worry too much. Texas still hadn't proved how much of that under-delivery had been caused by man's activities in New Mexico and until the Texans proved that, there was still no Compact liability.

That was, until the Supreme Court appointed Charles J. Meyers the new special master appointed to replace the frustrated Breitenstein. Then all hell broke loose in Whizzer White's "dog of a lawsuit."

V

Meyers, a former dean of the Stanford law school and a Texan, took hold of *Texas v. New Mexico* and within four years had shaken it so hard that the suit and the Compact lost all of its hair and emerged as a bare, skinny, relatively lifeless and mechanical legal obligation.

Meyers was abrupt and smart and in a hurry. In reviewing the case that he had inherited, Meyers learned how long it had taken just to get to the point where the baseline "1947 condition" had been settled. He didn't have another ten years while Reynolds and the New Mexicans confused and struggled with the equally troublesome definition of New Mexico's obligation not to have caused the departures from the 1947 condition by depletions by man's activities. After all, sifting rainfall patterns might have caused the departures. Changes in the Pecos River channel might have caused the departure. Those Pecos River "vampires," salt cedars and other phreatophytes, might have caused the departures. Indeed something as mysterious as "the perturbations of nature," as Reynolds and his lieutenants repeatedly said, could have caused the departures and New Mexico wasn't responsible for those, whatever they were. The upstream state was responsible only for those depletions caused by man's, not nature's, activities and it was up to Texas to parse those in the lawsuit. But Meyers wasn't willing to wait while the Texans tried and the New Mexicans fought them every step of the way.

Instead, Special Master Meyers stood Steve Reynolds' and New Mexico's position about departures and "man's activities" on its head. New Mexico, he said, was right: It was responsible for only that part of the annual departures in Pecos River flow at the state line that were caused by man's activities, but under the Compact it was up to New Mexico, not Texas, to separate them. Reynolds had counted on Texas's inability to parse the departures and thought it would take them forever to do it, if they ever could. Now here came Meyers and said it was New Mexico's obligation. It was not only that New Mexico couldn't do it. New Mexico hadn't done it and it was too late now in the litigation to try. New Mexico was responsible for all of the annual 10,000 acre-foot shortfall at the Texas/New Mexico state line.

This legal sleight of hand by Special Master Meyers transformed the battle over the 1948 Pecos River Compact from an engineering and scientific one to a legal one. Now it was the legal presumption that determined how much of the departures from the 1947 condition New Mexico would be responsible for. It wasn't the history of the development of water along the Pecos in New Mexico that determined how much of the reduced flow New Mexico would be responsible for. It wasn't the evolving science of the effect of groundwater pumping on surface-water flows that would determine its responsibility. What had started as a Compact that reflected how nature actually operated had ended up in Meyers' hands as a Compact that reflected the way that law operated.

There were other, even more fundamental changes that the Meyers ruling brought to the Pecos River Compact. Remember that the 1948 Pecos River Compact fundamentally started out as a depletion-based compact, one that favored upstream states like New Mexico. In one fell swoop of his lawyerly hand, Meyers eliminated the depletion-based element of the Compact's apportionment provision and thereby converted it into a flow-based compact, one that made New Mexico responsible not only for depletions caused by man's activities but for all changes in the River from its condition in 1947. This change was a disaster for New Mexico.

Meyers compounded the disaster by using his new standard for assessing damages against New Mexico for its past failure to abide by his brand new definition of the Compact obligation as well as setting New Mexico's obligation in the future. Over the previous 30 years, Meyers ruled, New Mexico had shorted Texas by 300,000 acre-feet and it was going to have to pay Texas back as well as produce an additional 10,000 acre-feet of water at the state line every year in the future. The combined past and future obligations would decimate the robust Roswell farming community, a major contributor to New Mexico's otherwise anemic economy.

Meyers tried to apply the *coup de gras* to New Mexico when he finally ruled that New Mexico would have to pay for its past and future sins in water, not money. The Supreme Court saw what a disaster this would be and quickly reversed Meyer on this point, ruling that New Mexico could pay back its past debt in cash not water. How much? Well, one University of Colorado economist estimated that Roswell farmers had made about a billion dollars (that's $1,000,000,000) out of the water that should have gone to Texas over the years. Other economists estimated that Texans had lost no more ten million ($10,000,000). Was the test of the money damages stemming from the breach of an interstate water compact to be measured by the benefit to the breaching state or the damage to the harmed state? No one knew at the time.

New Mexico legislators winced at the idea of coming up with a billion dollars to pay, God forbid, to Texans. State Engineer Steve Reynolds, increasingly wrought up about a disaster that was getting worse and increasingly worried that his long and distinguished career would end in a Pecos River catastrophe, fought with every means at his disposal to reduce the dollar amount. When the New Mexico State Legislature agreed to appropriate $14 million as damages and Texas agreed to accept that amount, everyone breathed a sigh of relief.

Reynolds himself carried the check to El Paso to deliver to the Texans. The Texas lawyer accepted delivery and told watching reporters that "New Mexicans have Steve Reynolds to thank for this." The unflappable Reynolds shot back: "I think that New Mexicans would be glad to buy a billion dollars worth of benefit for 14 million dollars any day." There in that unpleasant exchange forty years of New Mexico under-deliveries to Texas were bought and paid for and a lawsuit that had dragged on for more than 25 years came to an end not with a bang but a multi-million dollar whimper.

In the wider world of western water compacts, that end simply launched a cottage industry in interstate suits. Now for the first time it was clear that there were big damages to be collected for past violations of compacts and the lawyers jumped at the opportunity to collect big contingency fees. At the end of the 20[th] century and the beginning of the 21[st] Nebraska sued Colorado, Montana sued Wyoming, and every downstream state already compacted with an upstream state eyed the possibility of suing its upstream counterpart. At least *Texas v. New Mexico* set that possibility to rest on the Pecos River.

However, the resolution didn't answer the question of what New Mexico was to do in the future to bring itself into compliance with the future requirement that New Mexico produce 10,000 acre-feet of additional water annually at the state line. New Mexico could have enforced its existing laws to close the gap between what arrived at the Texas-New Mexico state line and what the Compact said should have gotten there. Instead the state chose a different path, one that cast a bright light on what the real water law of New Mexico is.

## VI

The New Mexico State Constitution sang the essence of New Mexico water in perfect pitch: The water belonged to the public and was subject to private appropriation for beneficial use. Public and private concerns merged in every water right. That's one reason why the state retained such an administrative interest in the most private New Mexico water right. For the better part of the 20th century, the private part of a water right held complete sway. Reynolds proposed through his administration to make private interests in water as secure, he said, "as the title to your house." Statements like that, of course, left out the significant public interest in private water. When it came time after his death in 1991 to figure out how New Mexico was going to produce at the Texas state line 10,000 acre-feet of additional Pecos River water per year, the contest between the public and private rights to water arose with a vengeance again.

In that context, three paths were open to New Mexico. In what was to many the most Draconian avenue to future compliance, the state could have taken the water from wherever it could find it and simply ordered that it be delivered to the Texas state line to satisfy the Compact. A 1920s compact covering the La Plata River running between southern Colorado and northwestern New Mexico authorized the two states to do this and the Supreme Court had approved. If New Mexico had selected this alternative, it could simply have directed the Carlsbad Irrigation District to dump the water it stored near the state line and send it to Texas even though under state law the CID had the senior rights to the river.

In the second alternative, New Mexico could have actively administered the Pecos River rights under state law and gotten the additional water to the state line using the state law mechanisms available to it. This route would have involved enforcing priorities under state law. The wet water needed to satisfy under-deliveries to Texas would have come out of the junior groundwater rights in Roswell, just as the CID had insisted in its standing 1976 priority call. At least with respect to surface water, Colorado had used this state law technique to bring itself into compliance with the Rio Grande Compact. Either of these established ways of dealing with Compact shortages wouldn't have cost the state a dime for water itself, although they both would have required an army to enforce and a cadre of engineers to even come close to making it work.

But the State of New Mexico couldn't bring itself to elect either of these established ways to bring itself into compliance with the new Supreme Court vision of the Pecos River Compact. Instead New Mexico, as always so far from Heaven and so close to Texas, took a third route, the legislative one. In a complex, long-term deal, the state agreed to reduce uses in New Mexico and to provide more water for Texas at the state line by itself buying from willing sellers and retiring from private use enough water rights to satisfy the state line demand. A certain larger percentage of those acres had to come from Roswell, the Legislature said, than had to come from Carlsbad. In addition, to make sure that wet water actually got there, some of the private rights to be purchased actually would be transferred to state wells. Exercising these purchased private rights, the new state wells could quickly get water to the state line and even guaranteed the CID some of the new state groundwater.

In this plan you could see the rejection of the basic scheme of New Mexico water rights. The risk of system shortage was an integral part of every prior appropriation water right and you didn't get paid for it if the water wasn't there. But the legislative solution treated every Pecos River right to water as perfect.

The state solved the problem of the public interest in water by turning itself into a private actor. Like a private party, it bought the water rights that it needed. A few it retired; most it transferred to three new state-owned well fields whose pumps would withdraw water if needed for the Compact. This $21^{st}$ century solution was an ironic and distorted reflection of the one that the deceased Steve Reynolds had proposed twenty-five years before in 1976. Those state wells would have taken the interconnected water and made Roswell pay for it. These state wells would take the water and the State of New Mexico would pay for it, a lot.

The state would pay more than $100 million in state funds for this solution. Every public entity agreed including the state, the CID, and the PVACD. But the private parties had to sign off as well. And the Tracy family from Carlsbad, a family that had been at the center of every water controversy in southeastern New Mexico for more than 100 years, wasn't about to let the Roswell farmers whom they had been trying to bring to account for a century off the hook so easily. They protested in what was to be the final legal challenge to the aftermath of *Texas v. New Mexico*.

## VII

They gathered in the make shift courtroom of the New Mexico Court of Appeals in Santa Fe one spring afternoon. The place was full of lawyers and farmers most of whom were old enough to have been involved in the Pecos River troubles at least since *Texas v. New Mexico* began in 1974 if not long before. Arrayed on one side, in support of the legislative solution to the Compact problem were, among others, Fred Henninghausen, who had started as the district engineer for the Roswell Office of the State Engineer and only later became a lawyer. Beside him sat Richard Simms who had been the lawyer for State Engineer Steve Reynolds for fifteen years before he'd moved on to a more lucrative private practice.

On the other side of the room, opposing these formidable icons of New Mexico water law, sat Louise Tracy, a couple of her brothers, and their lawyer, the late Paul Bloom. Louise Tracy was the granddaughter of Francis Tracy, the driving force behind the Carlsbad irrigation project since the turn of the 20$^{th}$ century and the daughter of Francis Tracy, Jr., the longtime head of the Carlsbad Irrigation District in the mid 20$^{th}$ century. The Tracys had spent generations fighting for their rightful share of Pecos River water and they weren't about to give up their rights in the swamp of this legislative deal.

The Tracy's lawyer, Paul Bloom, had been Steve Reynolds' right-hand man for almost twenty years in Reynolds' heyday in the 1960s and 1970s. The two had been so close that Reynolds felt betrayed when Bloom left the Office of the State Engineer and started to represent the City of El Paso, Texas in its effort to take New Mexico groundwater across the Rio Grande border to Texas. Reynolds never forgave Bloom for the perfidy. Now here was Paul Bloom again, trying to sink a deal that a new State Engineer had engineered and blessed.

On this spring day, 2006, before the New Mexico Court of Appeals, the courtroom seethed with the slightly decrepit energy of old, old battles, never fully resolved and the aging men who had fought them for decades or longer. The three judges from the Court of Appeals had not participated in the war and could only guess at the depth of the audiences' contrarian devotion to this ancient cause.

Paul Bloom, frumpled, old, but with the same intellectual elegance that he'd possessed since his University of Chicago days, opened for Louise Tracy. He offered to the Court a threnody for the principles of the basic New Mexico law of prior appropriation that this Pecos River settlement violated. Principal among these was the fact, according to Bloom, that priority enforcement was the only way sanctioned by the State Constitution and the Pecos River Compact to make up the shortages to the CID in New Mexico and Texas beyond. Anything other than that, including the public buying rights and drilling special wells, violated the Constitution, the Compact, and the ancient prior rights of the Tracys, that everybody had recognized for 100 years and nobody would enforce. "This shame," said Paul Bloom, "was a travesty of the legal system of prior appropriation that had been the law of New Mexico for hundreds of years."

In response, the CID's own lawyer, Steve Hernandez, the only lawyer in the house who had never worked for the State Engineer, took a different and simpler tack. "The law of New Mexico prior appropriation," Hernandez told the Court of Appeals panel, "is what the Legislature says it is, especially when it comes to the State's interstate obligations. The Legislature has spoken and you must uphold their decision."

The Court of Appeals panel took it all in. In the end, in a lengthy opinion issued in November 2006, the Court of Appeals sided with the PVACD-CID-State triumvirate and upheld the legislative solution that the Tracys hated. The Tracys, exhausted, did not appeal.

And there *Texas v. New Mexico* finally ended. The United States Supreme Court had re-written the Pecos River Compact. New Mexico had quietly and with judicial approval re-written its basic state law in response. Still the River itself and its basic legal obligations had come out looking better. River flows, including depletions, apparently were much closer to being in balance. Enough water was reaching Texas to satisfy its newly defined compact rights. The CID could satisfy the water rights of its constituents. And the remaining Roswell wells, always the most junior and the most productive, seemed to be in the clear even if those wells seemed to deplete more and more of the River base flows as they continued to pump. The transformed Pecos River Compact now in effect guaranteed to Texas half of the flood flows originating in New Mexico, even though it never said so, and this seemed fair to the River and all the parties.

# 6

# Albuquerque v. Reynolds: Conjunctive Use and Municipal Water Supply

## Jay F. Stein

*Albuquerque v. Reynolds* is significant both for its trailblazing principles unifying the administration of surface water and groundwater, as well as for prompting new approaches to the management of water rights in the State of New Mexico and other western states. This 1962 case, arguably the most important water case decided by a state court, established the principle of conjunctive use and management of hydrologically interrelated surface water and groundwater. At the time, this was a revolutionary concept. Today, the conjunctive use of surface and groundwater to exploit the advantages of both sources by diverting groundwater when the surface water is unavailable (in times of drought for example) is the cornerstone of sustainable water management. Conjunctive use has been adopted throughout the western United States.

In New Mexico, *Albuquerque v. Reynolds* provided the basis for long-term municipal water supply on the Rio Grande, exemplified by the needs of the City of Albuquerque. The case paved the way for the issuance of San Juan-Chama contracts for municipal, agricultural, and tribal interests on the Rio Grande, which would seek to augment their native Rio Grande water supplies with imported water from the Colorado River system. The decision led to the *de novo* review of state engineer decisions over water rights applications by the district courts. It cemented the career of Steve Reynolds as New Mexico's water czar.

All this was hammered out in water wars fought in the courts, debated in the legislature and in the halls of Congress, and reported by the press on page one. At stake was the Rio Grande, New Mexico's principal artery and the source of water for irrigators and its largest cities. The trigger was an interstate lawsuit brought by Texas against New Mexico alleging that New Mexico and the Middle Rio Grande Conservancy District (MRGCD) had violated provisions of the Rio Grande Compact, signed into law on May 31, 1939. The compact restricted the storage of native Rio Grande water in upstream reservoirs constructed after 1929. Texas alleged that New Mexico and MRGCD had violated these compact provisions, thereby shorting Texas water that it was owed. *Texas v. New Mexico* turned the spotlight onto New Mexico's administration of the Rio Grande and raised the need for policy that enabled the State to comply with its interstate obligations while conjunctively using interrelated ground and surface water for development in New Mexico. How this was accomplished by State Engineer Steve Reynolds, who used the device of declaring the Rio Grande Underground Water Basin as the means for providing an administration of water rights that both protected the surface flows and enabled groundwater development for municipal supply to proceed, is the story of *Albuquerque v. Reynolds*.

*Texas v. New Mexico*

In 1951, New Mexico was sued by Texas in the Supreme Court of the United States.

Texas alleged that New Mexico had violated the Rio Grande Compact. Texas attacked the storage practices of the MRGCD for El Vado Reservoir. This implicated the "prior and paramount rights" of six pueblo tribes, namely, the Pueblos of Cochiti, Isleta, Sandia, San Felipe, Santa Ana, and Santa Domingo.

In 1931, the state engineer had issued Permit No. 0620, which granted MRGCD the right to "change points of diversion" of 70-plus ditches to the present points of diversion "in accordance with the Official Plan of the Middle Rio Grande Conservancy District." In 1930 the state engineer had issued Permit No. 1690, which authorized the construction of El Vado Reservoir on the Rio Chama, the principal tributary to the Rio Grande. El Vado Reservoir, with a capacity of some 198,000 acre-feet of water, was constructed to store native Rio Grande water for use on about 80,000 acres claimed as irrigated by MRGCD. Permit No. 0620 referenced approximately 80,000 acres of irrigated land. Supply of these acres was intended in part to be from surface water, released when needed from El Vado Reservoir.

There was no corresponding regulation of groundwater use and development in the adjacent aquifer. In the early 1930s, Albuquerque was a city of about 30,000, continuing to develop a municipal supply from groundwater wells that it had begun to drill decades earlier.

Ratified in 1939, the Rio Grande Compact apportioned the main stem of the Rio Grande from its headwaters above the San Luis Valley in Colorado among the states of Colorado, New Mexico, and Texas. The Rio Grande Compact was signed "to remove all causes of present and future controversy among the states and between citizens of one of these states, and citizens of another state with respect to the use of the waters of the Rio Grande above Ft. Quitman, Texas, and being moved by considerations of interstate comity, and for the purpose of effecting an equitable apportionment of such waters. . . ." Rio Grande water delivered by Colorado and New Mexico is stored in Elephant Butte and Caballo Reservoirs.

The Rio Grande Compact apportions surface water. Articles III and IV, the apportionment provisions of the compact, require the upstream states of Colorado and New Mexico to deliver a percentage of the inflow recorded at gauges near Lobatos, Colorado, and at Otowi, New Mexico, downstream for storage in Elephant Butte Reservoir. "Usable water" in storage in Elephant Butte Reservoir has historically been divided 57 percent to Elephant Butte Irrigation District and 43 percent to El Paso Water Improvement District No. 1, the two components of the Rio Grande Project. The division of water between the two districts is made by contracts with the Bureau of Reclamation.

The key to the apportionment made by the Rio Grande Compact is that natural flows are required to be delivered downstream based on flow conditions that vary. Accordingly, storage in reservoirs upstream of Elephant Butte Reservoir is restricted in dry periods. Articles VII and VIII are the storage provisions of the Rio Grande Compact. Article VII states that "[n]either Colorado nor New Mexico shall increase the amount of water in storage in reservoirs constructed after 1929 whenever there is less than 400,000 acre-feet of usable water in project storage. . . ." Article VIII provides that the Commissioner for Texas may demand of Colorado and New Mexico, and the Commissioner for New Mexico may demand of Colorado, "the release of water from storage reservoirs constructed after 1929 to the amount of the accrued debits of Colorado and New Mexico, respectively . . . sufficient to bring the quantity of usable water in project storage to 600,000 acre-feet by March 1 . . . that a normal release of 790,000 acre-feet may be made from project storage in that year."

The apportionment made by the Rio Grande Compact is flexible. Unlike other interstate compacts to which New Mexico is a party, the Rio Grande Compact does not impose a rigid delivery obligation on the upstream states. Instead, within limits, Colorado and New Mexico may accrue debits and credits in their delivery obligations. Under Article VI, New Mexico may accrue up to 200,000 acre-feet of debits.

In its lawsuit, Texas alleged that beginning in 1948 New Mexico and the Middle Rio Grande Conservancy District had violated the storage provisions of the compact by retaining water in El Vado Reservoir for use within MRGCD that should have been released for use in Texas. In defending the suit, New Mexico contended that the United States was an indispensible party that had not consented to waive its sovereign immunity to suit, and thus the case must be dismissed. New Mexico's argument rested principally on the United States' trust responsibility for Pueblo tribes with an interest in water stored in El Vado Reservoir. New Mexico's argument was that the United States' interests were essential for a just determination of the issues raised by Texas, but that because the United States did not consent to be joined to the litigation, the case could not go forward in its absence.

The special master appointed by the Supreme Court to take evidence in the case concluded that the United States was indispensible with respect to the Pueblos' interests stored in El Vado Reservoir. He reasoned that the United States asserted rights to some 8,847 acre-feet of water with storage rights in El Vado reservoir with priority over the provisions of the Rio Grande Compact.

With a decision based on the legal principle of indispensable parties, the key substantive issues were not decided. But the need for administration that facilitated compliance with the compact, as well as the development of water resources for New Mexico's economy along the Rio Grande was clear.

## Declaration of the Rio Grande Underground Water Basin

On November 30, 1956, the headline in the *Albuquerque Journal* screamed: "Albuquerque Area Water Basin Established."

The accompanying article described "[a]n area roughly 280 miles in length and up to 25 or 30 miles wide in spots, extending from Elephant Butte Dam to the Colorado line, Thursday was proclaimed 'the Rio Grande underground water basin.'" The article went on to state: "[T]he declaration by State Engineer S.E. Reynolds after extensive and highly secret planning is expected ultimately to increase the water supply prospect for Albuquerque's municipal and industrial uses, Reynolds said." Three of the state engineer's five criteria for declaring the basin provided for "increase of beneficial use of water in the Rio Grande Valley over a number of decades," encouragement of "industrial development," and "maximum utilization of the underground water reservoir." At a press conference, the state engineer explained that "the City [of Albuquerque] may increase its own water rights by buying up existing rights and 'drying up' the land which those rights have been irrigating. Thus municipal-industrial usage could increase at the expense of retiring a proportion of the land formerly under irrigation." The *Journal*'s headline the next day confirmed: "Engineer Reynolds Visualizes Increased Basin Water Usage."

Under New Mexico law, the state engineer has jurisdiction over all surface water by virtue of the Surface Water Code of 1907. But the law contained an anomaly. It did not expressly encompass groundwater. That omission was rectified in 1931 with the enactment of the state's Groundwater Code, granting the state engineer jurisdiction over groundwater within "declared" underground water basins with "reasonably ascertainable boundaries" over which he asserts jurisdiction through the issuance of a "declaration" or order from his office, prompted by the need to regulate groundwater development. Outside the boundaries of a declared underground water basin, water users could initiate groundwater development, or transfer groundwater rights, without application to the state engineer. Not so once a basin has been declared. Then water users are required to apply to the state engineer for permits for new appropriations of water and for permits to transfer the points of diversion, or the purpose or place of use, of existing water rights. Applications are required to be noticed to permit protestants to appear at state engineer hearings. Permits are granted provided that the other rights are not "impaired," a standard kept deliberately vague to suit the needs of individual applications that are to be evaluated on their own merits. Other approval criteria require that the application not be contrary to the conservation of water within the state or detrimental to the public welfare of the state.

Between September and November of 1956, the state engineer, together with key members of his administrative staff, including Jim Williams, who would be transferred from the Roswell State Engineer District Office to open the Albuquerque District Office, worked to develop criteria to declare and administer the Rio Grande Underground Water Basin. New Mexico had accrued debits beyond those permitted by the compact. The impact of increased groundwater pumping on the surface flows of the Rio Grande was a growing concern.

In declaring the Rio Grande Underground Water Basin, the state engineer recognized the hydrologic connection between the surface flows of the Rio Grande and groundwater within the basin. Determining that new appropriations of groundwater would affect surface flows, his goal was to require that all future groundwater permits "offset" the effects of groundwater pumping. Because the timing of groundwater impacts upon the surface flows could be delayed, appropriators would be able to acquire offsets when their pumping was manifested on the surface flows of the Rio Grande.

Accordingly, the principles animating the declaration of the Rio Grande Basin were to provide for compliance with the Rio Grande Compact, and to facilitate the use of groundwater from the aquifer for municipal supply so long as the effects of groundwater pumping on the river were offset. The primary administrative goal of the declaration of the basin was to allow the continued development and use of the groundwater resource, facilitated through the use of offsets, which kept the river whole.

There were numerous detractors. Among them was the Albuquerque City Commission. Its Chairman, Maurice Sanchez, believed that declaration of the basin would cripple the economic future of Albuquerque. MRGCD was also concerned. On December 19, 1956, the *Albuquerque Journal* reported that following discussions on the new district with State Engineer Reynolds and officials of the City, Sanchez declared that he was "opposed to the district." He explained: "I do believe very strongly this order already has done irreparable damage to the city and its industrial expansion. . . ." On January 4, 1957, the *Albuquerque Journal* reported under the heading: "Underground Water District Draws Fire," that MRGCD attorney Martin Threet claimed that Reynolds had exceeded his authority.

The City filed suit. It unsuccessfully sought to stop enforcement of the regulations for the Rio Grande Underground Water Basin. The *Albuquerque Journal* reported on May 30, 1957, under the headline: "Judge Turns Down City's Plea in Case on Rio Grande Basin," the court's two-fold reasoning. The State had not consented to the suit, which was otherwise barred under the doctrine of sovereign immunity, which protected the State from suit where it had not consented to be sued, and the City had not exhausted "all of its administrative procedures." The exhaustion doctrine is a legal rule that requires litigants to have made their case before an administrative agency prior to seeking judicial relief in the courts. The court's approach to dismissing the case was accordingly based on "threshold" questions related to Albuquerque's ability to bring the suit.

In the case of *State of New Mexico v. Myers and Hoard*, two defendants, John W. Myers and well-driller E.T. Hoard, were convicted of drilling or deepening a well on Myers' property within the newly declared underground water basin without a permit. Myers' theory was that declaration of the Middle Rio Grande Underground Water Basin from the Colorado state line to Elephant Butte Dam "'having reasonable ascertainable boundaries' is absurd on its face." The New Mexico Supreme Court disagreed, concluding that "in view of the state of the record in this case we must presume that the action of the state engineer is correct." The court gave deference to the decision of the state engineer, which it viewed as a legitimate exercise of police power.

*Albuquerque v. Reynolds*

The City did not see the need for state engineer jurisdiction, or the state engineer's ability to require the City to obtain offsets for future groundwater pumping. Nevertheless, the City filed four applications for new appropriations of water within the newly declared underground water basin. Each application sought the appropriation of 1,500 acre-feet water per year to be pumped for municipal water supply. Each application referred to and incorporated a letter of transmittal in which the City stated that its claim as successor to the Pueblo de Albuquerque y San Francisco Xavier, founded no later than 1706, granted it an expanding right to the use of all waters, both ground and surface, within its limits under the Pueblo Water Rights Doctrine. This theory had been prompted by decisions in California for the City of Los Angeles and a district court decision in New Mexico. In 1959, the Pueblo Water Rights Doctrine was recognized in New Mexico for the City of Las Vegas in the case of *Cartwright v. Public Service Company of New Mexico*. The Pueblo Water Rights Doctrine has since been discontinued in New Mexico. Secondly, the City claimed the right to undertake diversions necessary for municipal supply under the Territorial laws of 1884 providing for a paramount right for municipalities to take water, albeit in the format of a condemnation proceeding when other rights were impacted.

Following hearing on the applications, the state engineer issued an order that the proposed appropriations would "impair existing rights to the use of the waters of the Rio Grande and that the city had refused to take the steps required by him to offset the adverse effect upon the rights of such users." The City's applications were denied.

The City appealed. Trial was held in the district court for the Second Judicial District in Bernalillo County. At trial, new evidence concerning the Pueblo Water Rights Doctrine invoked by the City was received. The court entered its judgment granting the City an "absolute and unconditional right" to appropriate water under this doctrine. The court ruled that the City was not required to comply with offset requirements sought to be imposed as conditions of approval by the state engineer because the City's Pueblo rights were vested prior to 1907.

The state engineer appealed the district court's ruling to the New Mexico Supreme Court, which reversed the district court. The supreme court held that the jurisdiction of the district court on appeal was no broader than that of the state engineer during the administrative hearing. Accordingly, the district court could not have received the evidence offered by the City of Albuquerque on the Pueblo Water Rights Doctrine. On the issue of offsets, the court's analysis focused on the groundwater appropriations statute and identified two threshold questions: (i) whether there were unappropriated waters; and (ii) whether the taking of such waters would impair existing water rights from such source.

It was here that the state engineer's key hydrologic finding that the surface and groundwater sources were interrelated came into play. The court had found that the City was subject to the appropriation statutes. The court held:

> . . . underground waters, in the area where the city proposed to drill its wells, contribute substantially to the flow of the Rio Grande, thus constituting a part of the source of the stream flow. The state engineer and the district court each found that the granting of the applications would impair existing rights to the use of the surface waters of the Rio Grande.

For its part the City asked whether the state engineer had the power and authority to interrelate the same as a matter of law so as to require the retirement of surface water rights as a condition precedent to the appropriation of groundwater.

The court focused on law in the western states where it "has always been the law that a prior appropriator from a stream may enjoin one from obstructing or taking waters from an underground source which would otherwise reach the stream and which are necessary to serve the stream appropriators' prior right. [citations omitted]" The court found:

> . . . it would indeed be anomalous for the legislature to enact laws designed to permit water, which would otherwise reach the stream in substantial quantities, to be withdrawn by pumps and thereby attempt to deprive the prior appropriators of their vested rights.

Having found the legislative authority in the state engineer to deny the City's applications on grounds of impairment, the court turned to the argument that the state engineer exceeded his power and jurisdiction by establishing and promoting rules and regulations requiring the retirement of surface water rights as a condition to new appropriations of groundwater in the Rio Grande Underground Water Basin. In addressing this argument, the court answered: "[I]f we assume, as we must, from the findings made by the state engineer and also by the district court that the underground waters in question cannot be taken without impairment to the rights of the river appropriators, even though there are unappropriated underground waters in the basin, then it would seem to follow that some method should be devised, if possible, whereby the available unappropriated water can be put to beneficial use." The court concluded: "[H]aving the statutory power and duty to prohibit the taking, by denying the applications in toto if necessary to protect existing rights, the state engineer has reasonably exercised his power by imposing suitable conditions so as to permit such taking as will not result in impairment."

## The Legacy of *Albuquerque v. Reynolds*

### Conjunctive Use

The principal legacy of *Albuquerque v. Reynolds* was establishment of conjunctive management and use of hydrologically connected surface and groundwater. *Albuquerque v. Reynolds* filled a void in the western states—recognition of the hydrologic connection between interconnected surface and groundwater, and the need to account for both conjunctively in water rights administration. In addressing this need, *Albuquerque v. Reynolds* tied the separate surface and groundwater codes together and expanded the authority of the state engineer to treat both as a whole. The same principle applies today to the conjunctive *use* of groundwater and *imported* surface water. Today, the conjunctive use of groundwater and imported surface water is a tool that promotes the sustainability of the aquifer while relieving the effects of pumping on native surface water and groundwater in storage.

Imported water

The decision in *Albuquerque v. Reynolds*, which seemed to limit the City's access to groundwater, led to the City's acquisition of imported water from the San Juan-Chama Project. The San Juan-Chama Project is a trans-basin water diversion that takes water out of the San Juan tributaries to the Colorado River in Colorado for importation and municipal and industrial use in the Rio Grande Basin in New Mexico. The project was authorized by Congress in 1962. The purpose of the San Juan-Chama Project is to furnish a water supply to the middle Rio Grande Valley for municipal, industrial, domestic, and agricultural purposes utilizing the State of New Mexico's apportionment of Colorado River water in the Upper Colorado River Basin Compact.

The San Juan-Chama Project diverts water from three tributaries to the San Juan River, a tributary of the Colorado River: the Navajo, Little Navajo, and Blanca Rivers for storage in Heron Reservoir on the Rio Chama. New Mexico contractors take delivery at the outlet works of Heron Reservoir. Pursuant to the authorizing legislation, uses for New Mexico's San Juan-Chama water were established by the New Mexico Interstate Stream Commission and adopted by the secretary of the interior.

On March 4, 1953, New Mexico Governor Mechem requested the secretary of the interior to study and make feasibility reports on projects in the state for using imported San Juan-Chama water. Governor Mechem's letter of March 4, 1953, identified the State's intent with respect to San Juan-Chama water. Governor Mechem wrote:

> I, therefore, respectfully request that you direct the proper agencies of the Department of the Interior to study and make feasibility reports on projects in New Mexico utilizing San Juan River water as follows:
> (1) A Shiprock Project***
> (2) A San Juan-Chama project to transport water from tributaries of the San Juan River in Colorado to the Rio Grande Basin in New Mexico by means of a transmountain diversion.
>
> This project should be investigated in accordance with the expressed policy of the State that such transmountain water shall be

used primarily for domestic, municipal, and industrial supplies, and for supplemental use on existing projects with deficient supplies and that preference in the irrigation of new lands shall be given to inbasin projects.

In the "Plan for Development" of the San Juan-Chama Project of November, 1955, express mention was made of supply requirements for the City of Albuquerque:

> The requirements of the Albuquerque metropolitan area, the most practicable to provide, are expected to be more than 90,000 acre-feet by the year 1990. On the assumption that 40,000 acre-feet or more could continue to be obtained from the present source, the city of Albuquerque, through a resolution, requested that an allotment to the city of 50,000 acre-feet of water by 1990 be considered when the allocation of San Juan River water is made. Satisfaction of this request would involve a requirement of 55,800 acre-feet of San Juan River water, as the losses involved in storage of the imported water and its transmission through the channels of the Rio Chama and Rio Grande to Albuquerque would total 5,800 acre-feet. Prior to passage of the resolution requesting 50,000 acre-feet of San Juan-Chama project water, the city evidenced its requirements for additional water supplies by filing an application with the State engineer for 150,000 acre-feet of San Juan River water.

After the decision of *Albuquerque v. Reynolds* was final in 1963, Albuquerque turned in earnest to importing San Juan-Chama water from the Colorado River Basin to provide for the future growth anticipated within 40 years. The City Commission signed its San Juan-Chama contract with the Department of Interior on June 25, 1963, granting it 52,000 acre-feet of imported Colorado River water. A significant feature of this water was that as imported water, it is owned by the City and subject to full consumptive use. This contract was subsequently amended in 1965 when the City's allotment was reduced to 48,200 acre-feet to provide water to the state for Jemez Canyon Reservoir. The City has a perpetual contract for the consumptive use of 48,200 acre-feet annually. After release, Albuquerque's water is stored in Abiquiu Reservoir where the City has a storage account containing 170,900 acre-feet of water.

Albuquerque's example was followed by other cities on the Rio Grande including Santa Fe, Belen, Los Lunas, and others; by the Middle Rio Grande Conservancy District; and by certain tribes including San Juan Pueblo (now the Pueblo of Ohkay Owingeh).

**Trial *de novo***

A final irony emerged in 1967. *Albuquerque v. Reynolds* was decided by the district court under the "arbitrary and capricious" standard in which deference was given to the findings of the state engineer from the administrative hearing. In 1967, legislation was introduced and signed by the governor under which appeals of state engineer decisions to the district courts were heard as trials *de novo* as cases originally docketed in the district court in which new evidence could be received. This process was finalized by a constitutional amendment in 1969. Had it been in place in 1960, the City's evidence with respect to the Pueblo Water Rights Doctrine would not have been stricken by the New Mexico Supreme Court on appeal. Accordingly, while *Albuquerque v. Reynolds* expanded the state engineer's authority by giving him the ability to consider impairment to surface water rights when considering the appropriation of hydrologically connected groundwater, it led to a contraction of the weight given his administrative decisions. Appeals to the district court that had previously been governed by the "arbitrary and capricious" standard, which limited appeals to a record review, were replaced by the *de novo* standard that opened the district court proceedings to the determination of applications as if originally filled as district court actions.

# 7

# Las Vegas, New Mexico: The Rise and Fall of the Pueblo Water Rights Doctrine

James C. Brockmann and Eluid L. Martinez

The essence of the Pueblo Water Rights Doctrine is that a municipality that can trace its origins to a Spanish or Mexican pueblo grant has a prior, paramount, and expanding right to all waters of non-navigable streams flowing through or by the pueblo to the extent necessary to serve its present and future growth. The doctrine relates to Spanish or Mexican pueblos, not Native American pueblos.

The effect of the Pueblo Water Rights Doctrine is to give municipalities that can comply with all elements of the doctrine an ever-expanding right to appropriate all surface water and inter-related groundwater from non-navigable streams flowing through or by the pueblo, eliminating the need to continually obtain new appropriations or transfers of additional water rights to serve larger populations over time. Pueblo water rights are not favored by non-municipal water users, particularly acequias, because recognition of such a right allows vested water rights to be taken over time as a municipality increases its use of a stream or inter-related groundwater. The New Mexico state engineer has not favored the recognition of pueblo water rights, in part, because they are difficult to administer in a system of prior appropriation because the quantity of the water right is never concretely determined.

New Mexico's relationship with the Pueblo Water Rights Doctrine has had a mixed history. As told in the story of the City of Las Vegas' water rights, the New Mexico Supreme Court originally held that the Pueblo Water Rights Doctrine was legally valid in New Mexico and that Las Vegas had a pueblo water right based upon the facts of that case. Over 40 years later, the New

Mexico Supreme Court reversed its earlier decision and ruled that the Pueblo Water Rights Doctrine was not valid in New Mexico, changing the landscape significantly for Las Vegas and other potential successors to Spanish or Mexican colonization pueblos.

Not long after its adoption and use in California, the Pueblo Water Rights Doctrine was first raised in New Mexico in the early 1900s in the context of a dispute over the management and control of the Tularosa Community Ditch. Two rival organizations claimed the right to administer and control water rights and water use from the Tularosa Community Ditch, and one of the major issues was whether water rights could be sold and transferred outside the original town site. In *State ex rel. Community Ditches v. Tularosa Community Ditch*, 19 N.M. 352, 143 P. 207 (S.Ct. 1914), the New Mexico Supreme Court did not address the validity of the Pueblo Water Rights Doctrine, but found that because the Village of Tularosa was first settled in 1862, the municipality could not trace its origins to a pueblo of Spanish or Mexican descent. Rather, the town site grant was made by the federal government long after New Mexico became a part of the United States and therefore was governed by the laws of the granting sovereign and not an earlier sovereign. The court stated that "[w]hatever might have been the rights of the people of this settlement, had the land been acquired from the Mexican Government by grant, or otherwise, is of no consequence. The land having been acquired from the United States, after it passed under its jurisdiction and control, the grant would carry with it only such rights and privileges as were accorded by the laws of the United States." In short, under the facts of this case, the Village of Tularosa was founded long after the territory in which the village was located was acquired by the United States and had never been a Spanish or Mexican pueblo.

Approximately 20 years later, the City of Santa Fe through its domestic water provider, the New Mexico Power Company (collectively the "City of Santa Fe"), claimed water rights derived from the Pueblo Water Rights Doctrine. In *New Mexico Products Co. v. New Mexico Power Co.*, 42 N.M. 311, 77 P.2d 634 (S.Ct. 1937), the New Mexico Products Company claimed that it had senior water rights for irrigation and that those rights were being interfered with by the City of Santa Fe. The district court agreed with Santa Fe that the Pueblo Water Rights Doctrine was valid in New Mexico and that the City of Santa Fe "had the right—regardless of the prior appropriation and beneficial use by others—to take from the Santa Fe creek from time to time all the water that may be needed at such time for the use of the inhabitants of said city and for all municipal and public uses and purposes therein."

The New Mexico Supreme Court disagreed with the district court. In an earlier, unrelated case entitled *United States v. City of Santa Fe*, 165 U.S. 675 (1897), the United States Supreme Court had held that Santa Fe did not hold title to *land* pursuant to the Pueblo Water Rights Doctrine because there had never been an official grant from the Spanish king. The United States Supreme Court held that the occupancy of the pueblo by the Spanish military and governmental authorities, in and of itself and without an official grant, conferred no title on the inhabitants. Drawing on the United States Supreme Court's decision, the New Mexico Supreme Court held that it was settled law that no grant had ever been made by the Spanish king to the Villa de Santa Fe. Without a grant, the Villa de Santa Fe had no pueblo right, for either land or water rights. The New Mexico Supreme Court stated that it had found no law to suggest "that a mere colony of 'squatters' could acquire under the Spanish law this extraordinary power over the waters of an entire non-navigable stream known as 'pueblo right,' even though they were organized as a pueblo—which is the equivalent of the English word 'town'—with a full quota of officers."

Approximately 20 years after *New Mexico Products Company*, the Pueblo Water Rights Doctrine was again the subject of litigation, this time involving the City and Town of Las Vegas, New Mexico. This time the result was much different.

A number of water users on the Gallinas River filed suit against Public Service Company of New Mexico (PSC), claiming that PSC was trespassing by diverting their senior water rights. The plaintiffs sought an injunction and damages against PSC. PSC was a water company that provided municipal water service to two neighboring communities, the Town of Las Vegas and the City of Las Vegas (which later consolidated). The case was styled as *Cartwright et al. v. Public Service Co.*, 66 N.M. 64, 343 P.2d 654 (S.Ct. 1959) (*Cartwright*), and the Town of Las Vegas sought and was granted leave to intervene. Among other defenses, PSC claimed that the Town of Las Vegas had a prior and paramount water right to divert an unlimited amount of water from the Gallinas River to supply existing municipal needs and increased municipal demands in the future pursuant to the Pueblo Water Rights Doctrine, such rights being traced back to the Town's origins under Mexican law.

The district court made factual findings that: 1) the City and Town of Las Vegas were successors to the Mexican pueblo of Nuestra Senora de

Las Dolores de Las Vegas which was established on April 6, 1835, by the government of the Republic of Mexico; 2) the government of the Republic of Mexico made and approved a community colonization grant known as the Las Vegas Grant to the pueblo known as Nuestra Senora de Las Dolores de Las Vegas; 3) the Gallinas River ran through and beside the Mexican pueblo and was its sole source of water; 4) the territory within which the pueblo was located became part and under the jurisdiction of the United States of America by the Treaty of Guadalupe Hidalgo in 1848; 5) the United States of America confirmed the grant made by the Republic of Mexico to the Mexican pueblo and issued to the Town of Las Vegas a patent to the Las Vegas Grant; and 6) the laws of the Republic of Mexico in effect when the Mexican pueblo of Nuestra Senora de Las Dolores de Las Vegas was established, and continuing to exist when the Treaty of Guadalupe Hidalgo was signed, provided that Mexican colonization pueblos should have prior and paramount rights to the use of all of the water they needed from streams or rivers flowing through or near the pueblos for the use of the pueblos, including any amounts necessitated by growth in the future.

The district court concluded as a matter of law that the City and Town of Las Vegas, as successors to the Mexican pueblo Nuestra Senora de Las Dolores de Las Vegas, had a right to divert and use as much of the waters of the Gallinas River as was necessary for their use with a priority date of 1835, which rights were prior and paramount to any rights of the plaintiffs in the case. The district court also concluded that PSC was acting on behalf of the City and Town under certain franchise agreements in diverting and distributing water for municipal use.

By a narrow 3-2 margin and over a vigorous dissent, the New Mexico Supreme Court upheld the decision of the district court and found the Pueblo Water Rights Doctrine to be valid in New Mexico. In addition, the court found that the City and Town of Las Vegas had met their factual burden of proof to establish such a right. Generally, the New Mexico Supreme Court found support for the Pueblo Water Rights Doctrine in several cases from California and a handful of legal treatises. More specifically, the court reasoned that the Pueblo Water Rights Doctrine made sense because when pueblos were initially provided with colonization grants, the area was largely, if not always, established before there was any other settlement of the surrounding area. As a practical matter, this meant that there had not been prior appropriations or use of waters from the streams, giving notice of the prior use of and entitlement to water. In addition, the court stated that there was nothing inconsistent with the Pueblo Water Rights Doctrine and the doctrine of prior appropriation and beneficial use.

To fully resolve this matter, the state supreme court had to resolve two other matters. First, the plaintiffs claimed that the City's and Town's water rights claims had been resolved by the Hope Decree entered in 1933, a prior stream adjudication conducted in federal court, and that those claims should not be relitigated in the *Cartwright* case. The court ruled that the predecessor to PSC had been a party to the Hope Decree, not the City or Town of Las Vegas, so the Hope Decree had no binding effect on the municipalities' pueblo water rights claims. Second, plaintiffs alleged that they had valid and superior title to the Las Vegas Grant by way of a grant initiated in 1821 to Luis Cabeza de Baca. Relying heavily on the fact that the United States Congress confirmed the Las Vegas Grant as a valid Mexican grant to the Town of Las Vegas, the court found the Mexican pueblo claim superior to the conflicting claim of the heirs of Luis Cabeza de Baca.

Judge William R. Federici wrote a lengthy dissent challenging nearly every aspect of the majority decision. Initially, he disagreed with the majority that the Hope Decree did not previously resolve the water rights claims of the City and Town of Las Vegas. In one of the most poignant statements in New Mexico water law, Judge Federici stated: "How in the name of recognized New Mexico water law and under the documentary evidence in this case can defendant Public Service Company now be heard to claim that it is entitled, if necessary, under an alleged Pueblo Doctrine, to take the whole of the water from the natural, upper public water courses of the applicable stream system?" Next, he analyzed the Pueblo Water Rights Doctrine, concluding that it should not be applied in New Mexico. With respect to the California precedent, Judge Federici stated that there "are many reasons why this Court should not follow the California cases which have engrained their strained interpretation of the Pueblo Doctrine to the many other apparent hybrid doctrines also existing in that state. . . ." Beside his opinion that the California cases misinterpreted Mexican and Spanish laws, he found that certain legislation in that state helped ensure Los Angeles a firm water supply that it was having problems securing strictly through judicial means. With respect to the treatises' description of the Pueblo Water Rights Doctrine relied on by the majority of the court, it was Judge Federici's opinion that the commentators were merely reciting the California decisions and not endorsing them.

Judge Federici also set forth facts that demonstrated that the Las Vegas Grant was settled with individuals making appropriations of water prior to the colonization grant of the Mexican pueblo Nuestra Senora de Las Dolores de Las Vegas, undercutting one of the stated rationales in favor of the doctrine. Importantly, he also stated his belief that the Pueblo Water Rights Doctrine was not consistent with the New Mexico law of prior appropriation and conflicted with the state constitution. He warned of vested water rights in the Rio Grande and Pecos River being lost to growing municipalities.

The dissent made clear that its views were not meant to deprive Las Vegas of necessary municipal water. Judge Federici stated that "municipalities do have a preferential right but such right is a preference as developed by the law of appropriation rather than preference as set out in the California cases. This court has repeatedly held that New Mexico follows the Colorado Doctrine. *The Colorado courts allow a preference but require compensation.*" (emphasis in original) In his view, the municipal power of eminent domain with respect to water rights was a critical component of water law in New Mexico that cut against the need for a pueblo water rights doctrine.

Two motions for rehearing were filed, with both motions being denied by split decision. In both instances, lengthy dissents were written expounding on arguments made in earlier dissents.

The legal recognition of the Pueblo Water Rights Doctrine in *Cartwright* was a significant victory for Las Vegas. It was also a windfall for any other New Mexico municipality that could prove it was a successor to a Spanish or Mexican pueblo that had received a colonization grant prior to the 1848 Treaty of Guadalupe Hidalgo. Legal recognition of this doctrine for qualifying municipalities equated to an ever-expanding prior and paramount municipal water right that could never be curtailed by another appropriator and implementation of the doctrine was seen as a threat to vested water rights. The legal activity related to the Pueblo Water Rights Doctrine was not confined to Las Vegas, but extended to the other New Mexico municipalities.

While *Cartwright* was pending, the City of Albuquerque filed four applications with the state engineer for a new appropriation of groundwater. Each application referred to and incorporated by reference a letter in which the City stated that by filing the application with the state engineer, it was not waiving or abandoning its claim that as the successor to the Pueblo de Albuquerque y San Francisco Xavier, founded not later than 1706, it had an absolute right to the use of all waters, both surface water and groundwater, within its limits for the use and benefit of its inhabitants. The state engineer did not rule on Albuquerque's pueblo water rights claim and denied the applications on other grounds.

The City appealed the state engineer's decision to the district court that found that Albuquerque had a pueblo water right. The district court held that the state engineer had no jurisdiction to impose conditions of approval on a permit issued to Albuquerque because the City, as the successor of the pueblo San Felipe de Albuquerque, had an absolute and unconditional right to divert and use as much of the surface water and groundwater of the Rio Grande as is necessary for its use. The New Mexico Supreme Court overruled the district court on appeal in *City of Albuquerque v. Reynolds*, 71 N.M. 428, 379 P.2d 73 (S.Ct. 1962), finding that Albuquerque was attempting to secure an adjudication of a pueblo water right without notice of any kind to other appropriators in the Rio Grande Basin. The state supreme court determined that the district court had no jurisdiction to consider and adjudicate the City's pueblo water right claim and that the issue was not properly before it. The New Mexico Supreme Court's decision in *Albuquerque v. Reynolds* was not a substantive decision on the merits, but rather a decision that the City must pursue its claim for a pueblo water right in a different venue.

By the time *Albuquerque v. Reynolds* was decided in 1962, just three years after *Cartwright*, there were all new justices on the supreme court. While Las Vegas' pueblo water rights claim was decided in the context of private parties competing for water from the Gallinas, the supreme court in *Albuquerque v. Reynolds* would not let Albuquerque's pueblo water rights claim be resolved without all potential claimants in the Rio Grande Basin. *Albuquerque v. Reynolds* can be seen as the earliest effort to back away from elements of the *Cartwright* decision.

While Las Vegas might have believed that it could rely on its pueblo water rights permanently into the future and the supreme court might have believed that it had legally resolved whether the Pueblo Water Rights Doctrine would be applicable in New Mexico, neither was true. The *Cartwright* decision was highly controversial. After the supreme court's decision, the majority of the commentators opined that the dissent was better reasoned than the majority opinion and that the court made both legal and factual mistakes in upholding a pueblo water right for Las Vegas. In addition, New Mexico's state engineer at the time, Steve Reynolds, and his successor, Eluid Martinez, argued that the Office of the State Engineer was not bound by *Cartwright* because the state engineer was not a party to the *Cartwright* litigation. Fortuitous for the New Mexico state engineer, another forum existed in which it could challenge the Pueblo Water Rights Doctrine, despite *Cartwright.*

The State of New Mexico through the state engineer had filed a suit to adjudicate all of the water rights to the Pecos River, including the Gallinas River. Whereas the earlier *Cartwright* decision only involved Las Vegas and individual appropriators on the Gallinas, this lawsuit involved a stream-wide adjudication that would define and quantify the water rights of Las Vegas as between Las Vegas, the State, and all water users on the Pecos River Stream System. The district court decided that it would carve out a sub-proceeding as part of the larger adjudication to resolve Las Vegas' water rights, a sub-proceeding that would involve Las Vegas, the state engineer, and all other water rights claimants that may want to present an *inter se* challenge to Las Vegas' water rights claims. In the late-1980s in the context of that sub-proceeding, both the state engineer and the City sought to avoid or narrow the issues for trial on Las Vegas' water rights so they filed cross-motions for summary judgment. The district court denied both motions for summary judgment and the matter was then heard on appeal by the New Mexico Court of Appeals.

In its motion for summary judgment, the State pointed out that prior to the 1933 Hope Decree, there had been an earlier lawsuit seeking to adjudicate surface flows of the Gallinas River that resulted in a 1922 Gallinas Decree. The 1922 Gallinas Decree had not recognized any pueblo water rights and arguments related to the decree had not been raised in the *Cartwright* case. In the appellate decision that was entitled *City of Las Vegas v. Oman*, 110 N.M. 425, 796 P.2d 1121 (Ct. App. 1990), the New Mexico Court of Appeals found that because the 1922 Gallinas Decree was inconsistent with the *Cartwright* case and the *Cartwright* case was resolved later in time, *Cartwright* controlled, and therefore it upheld the district court's denial of the State's motion for summary judgment.

In Las Vegas' motion for summary judgment, the City relied on the *Cartwright* case to argue that its pueblo water right must be recognized and that the only issue for trial was the quantification of that right. The court of appeals held that because the State and all other water users on the Gallinas were not parties in the *Cartwright* case, they were not bound by that decision. However, because the state supreme court held in *Cartwright* that the Pueblo Water Rights Doctrine was legally recognized in New Mexico, the legal doctrine of *stare decisis*, or the principle that precedent controls, applied. This principle required the district court and parties to the adjudication to recognize the existence of the Pueblo Water Rights Doctrine, that is, that a successor to a colonization grant is entitled to a pueblo water right.

That, however, was not sufficient to grant Las Vegas' motion for summary judgment. The court of appeals stated that because the *Cartwright* decision involved issues of both law and fact, the parties that did not participate in the *Cartwright* decision were not bound by it and could present factual evidence that could undercut Las Vegas' pueblo water rights claim. The district court had previously found that there were unresolved factual issues related to the existence, parameters, and ownership of the pueblo water right claimed by Las Vegas. The court of appeals agreed, finding that the district court may have to resolve issues that were not addressed in *Cartwright*. In addition, it stated that it was within the discretion of the district court to allow the parties to make a record upon which the Pueblo Water Rights Doctrine could be challenged in the New Mexico Supreme Court.

Rather than simply denying the motions for summary judgment to allow additional factual evidence to be taken at trial related to the pueblo water rights claim, the court of appeals recognized the controversial nature of the *Cartwright* decision and made clear that development of the record below could aid the supreme court should it decide to overrule *Cartwright*. The *Oman* court stated:

> [T]he district court may decide on remand to permit an adequate record to be developed so that ultimately the supreme court will be in a position to overrule *Cartwright I* if it chooses to do so. . . . (citation omitted) . . . We believe the supreme court's ability to develop case law will be best served if, when a district court makes a preliminary determination that the supreme court might reconsider one of its prior decisions, the district court is permitted to allow an offer of proof or otherwise permit the development of an adequate record. Thus, we hold that the district court is not precluded . . . from receiving evidence regarding the historical accuracy of the pueblo water rights doctrine, if the district court believes that the supreme court might reconsider the general principles announced in *Cartwright I*.

Accordingly, all motions for summary judgment were denied and the matter was remanded to the district court for trial on Las Vegas' water rights.

The district court declined to address the validity of the Pueblo Water

Rights Doctrine and considered itself bound by *Cartwright* to recognize the Pueblo Water Rights Doctrine. For appellate review, however, it did allow the parties to tender proof at trial.

Both the City and the State appealed the district court's decision and it landed back on the desk of the court of appeals. On appeal, the State claimed that the Pueblo Water Rights Doctrine was historically invalid, and Las Vegas claimed that the validity of the doctrine was resolved in *Cartwright* and was not before the court. This case was called *State ex rel. Martinez v. City of Las Vegas*, 118 N.M. 257, 880 P.2d 868 (Ct. App. 1994) and in it, the court of appeals analyzed four major issues. First, it examined whether it was required to follow and apply *Cartwright*. Second, it reviewed the historical validity of the Pueblo Water Rights Doctrine. Third, it addressed whether the Pueblo Water Rights Doctrine was compatible with New Mexico's law of prior appropriation. And fourth, the court considered the City's reliance on the Pueblo Water Rights Doctrine.

Initially, the court of appeals analyzed whether it was strictly bound by *Cartwright* to recognize the Pueblo Water Rights Doctrine as it had originally concluded it must do in *Oman*. This time, it reached a different conclusion. The court of appeals interpreted recent case law from the New Mexico Supreme Court to mean that that even though it was an intermediate appellate court and generally bound by decisions of the supreme court, there were certain circumstances in which it could choose not to follow supreme court precedent. With that conclusion, the court of appeals stated that it was not constrained to follow *Cartwright* and, in fact, was not going to be bound by that precedent.

Next, the court analyzed the historical validity of the Pueblo Water Rights Doctrine. The court noted that *Cartwright* had relied, in part, on treatises describing the Pueblo Water Rights Doctrine. The *Martinez* Court found that none of the treatises cited to any Spanish or Mexican legal authority, but rather, exclusively relied on the California cases establishing the doctrine. The court then reviewed the California cases that established the doctrine and found only faint evidence, at best, to support the proposition that pueblo water rights were impliedly granted to Mexican and Spanish colonization pueblos on their formation. The *Martinez* Court then summarized the "[c]onsiderable original and comprehensive research [that] has been done on the subject of Spanish and Mexican water law since *Cartwright I* was decided." Finding that scholars and historians agreed that giving a municipality an absolute preference to a river's water was not supported by the laws or practices of the Spanish or Mexican governments prior to 1848, the court concluded that the overwhelming weight of the scholarly evidence refuted the Pueblo Water Rights Doctrine. The *Martinez* Court stated that it was "convinced that the doctrine is historically invalid and that our supreme court would overrule *Cartwright I* if this appeal were before it."

Next, the court found that the Pueblo Water Rights Doctrine was fundamentally incompatible with New Mexico's law of prior appropriation. The *Martinez* Court pointed out how the Pueblo Water Rights Doctrine interfered with the state engineer's supervision and control of the public waters, was contrary to municipalities' 40-year planning horizon as established by the legislature, and could interfere with the administration of compacts on interstate streams and rivers.

Finally, the court of appeals addressed Las Vegas' claim that because it had relied on *Cartwright* for over 35 years, it was a "rule of property" that should not be disturbed. The City claimed that by relying on *Cartwright*, it had foregone the purchase of other water rights, invested in a pipeline from Storrie Lake, and diverted water from the Gallinas, actions it would not have done but for the *Cartwright* Court's recognition of its pueblo water right. Unswayed, the court of appeals found no such "rule of property" and no detrimental reliance by Las Vegas.

Summarizing its conclusions, the *Martinez* Court stated:

> The recent scholarship that has shown the pueblo rights doctrine to be historically invalid, combined with the practical difficulties of continuing to recognize the doctrine, convince us that our Supreme Court would overturn *Cartwright I* if this case were before it. Consequently, we decline to follow *Cartwright I*, and therefore hold that the City has no pueblo rights to water. We also hold that the City has not reasonably relied on *Cartwright I* for any use to which it has put water. We thus reverse the trial court's order and remand for further proceedings consistent with the opinion.

The most significant departure of the *Martinez* Court from the *Oman* Court was its decision as an intermediate appellate court to break from New Mexico Supreme Court precedent.

Not surprisingly, the *Martinez* Court's decision was appealed to the state supreme court. The decision on appeal was delayed at the request of the parties while they engaged in settlement negotiations that were ultimately unsuccessful. After several years, the matter was submitted to the court for resolution and the court resolved the matter in *State v. City of Las Vegas*, 2004-NMSC-009, 135 N.M. 375, 89 P.3d 47 (*Martinez II*).

Initially, the New Mexico Supreme Court made clear that the court of appeals was in fact bound by its precedent in *Cartwright*. In other cases, however, it had invited the lower appellate court to explain any reservations that it might have over the application of its legal precedent to the case below so that the state supreme court would be in a more informed position to decide whether to reassess prior case law. The *Martinez II* Court interpreted the lower court's opinion as expressing reservations over the Pueblo Water Rights Doctrine adopted in *Cartwright* and found that any attempt by the New Mexico Court of Appeals to actually overrule *Cartwright* to be harmless error.

On the merits of the Pueblo Water Rights Doctrine, the State argued that there was no historical basis for the Pueblo Water Rights Doctrine in Spanish or Mexican law. The *Martinez II* Court opined that while the Pueblo Water Rights Doctrine rested on a very narrow foundation, there was no historical evidence sufficient to justify overruling *Cartwright* on that basis. Despite its analysis related to the historical basis for the Pueblo Water Rights Doctrine, the *Martinez II* Court ultimately found this issue did not resolve the case. Instead, it stated that "because we conclude . . . that the pueblo rights doctrine is inconsistent with New Mexico law and not protected by the Treaty of Guadalupe Hidalgo, the historical validity of the pueblo rights doctrine is irrelevant to our determination that *Cartwright* must be overruled."

The *Martinez II* Court found the State's arguments that the Pueblo Water Rights Doctrine is incompatible with New Mexico water law and violates public policy much more convincing. In particular, the court found the perpetually expanding nature of the Pueblo Water Rights Doctrine conflicts with the fundamental principles of beneficial use that lie at the heart of New Mexico water law. For example, pursuant to state water law, water users have a reasonable time after the initiation of an appropriation to place the water to beneficial use. Municipalities are allowed additional time through a 40-year statutory planning period. Inconsistent with that principle is the Pueblo Water Rights Doctrine that allowed an unlimited time to place water to beneficial use. In addition, pursuant to the law of prior appropriation water rights can be lost through forfeiture, another principle that does not apply to pueblo water rights.

Because of the incompatibility of the Pueblo Water Rights Doctrine with the law of prior appropriation recognized in New Mexico, the *Martinez II* Court found that by "facilitating the underutilization of essential public waters, the pueblo right prevents the efficient, economic use of water that is necessary for survival in this arid region and upon which our entire system of water law is based." Accordingly, the New Mexico Supreme Court overruled *Cartwright*.

The *Martinez II* Court still needed to resolve two issues—first, whether it would apply its ruling retroactively to Las Vegas, and second, the nature of a water right that was initiated during colonization of a Mexican or Spanish pueblo as applied to other New Mexico municipalities.

With respect to the City of Las Vegas, the supreme court ruled that while the City does not possess a pueblo water right that expands indefinitely to meet its growing needs, it had to some extent relied on *Cartwright*, and to ameliorate any potentially harsh impacts on the City through a strict application of the *Martinez II* decision, the case would be remanded to the district court to determine an equitable remedy that would balance the City's reliance on *Cartwright* with other water users' reliance on New Mexico's established water law of prior appropriation.

With respect to other municipalities, instead of a pueblo water right, the *Martinez II* Court concluded that the water right acquired by a municipality under a colonization grant from an antecedent sovereign is recognized in New Mexico in the same manner as other municipal water rights. Said differently, under the doctrine of prior appropriation, a city's colonization grant can create a vested water right, but only to as much water as the pueblo placed to beneficial use within a reasonable time after the initiation of the appropriation, subject to factual proof.

Accordingly, 46 years after *Cartwright* recognized a prior and paramount expanding municipal water right for Las Vegas, the Pueblo Water Rights Doctrine was overturned, reaffirming the proposition that all municipal water rights must be quantified on the basis of the law of prior appropriation.

What happened to Las Vegas' pueblo water right? It has not yet been resolved. The state engineer and Las Vegas reached a tentative agreement on the elements of the City's water right that was initiated by the pueblo grant, only to have the district court rule that all claimants to the water in the Gallinas River Stream System needed to receive notice and have an opportunity to object. The logistics of that process, along with a failed attempt at mediation, have delayed resolution of the City's claim. Over 50 years after Las Vegas began its quest to obtain a pueblo water right, it continues its battle today, having won and then lost on recognition of the Pueblo Water Rights Doctrine, but still battling to obtain equitable relief for having ostensibly relied upon the doctrine for more than 45 years.

The practical effect of the *Martinez II* Court's decision is that New Mexico municipalities must obtain new water supplies to meet the needs of expanding populations by new appropriations, the purchase and transfer of existing water rights, water conservation, or water supply projects that import additional water into their water supply systems. If a municipality needs to acquire and transfer existing water rights from a non-navigable stream or related aquifer flowing through or by the municipality, it must compensate the owners of those vested water rights.

# 8

# The Middle Rio Grande Minnow Wars

Charles T. DuMars

*The views contained in this chapter are the author's and not those of any client or political subdivision.*

The Prior Appropriation Doctrine for water allocation has been the law in New Mexico since well before statehood. That Doctrine, coupled with another principle imbedded in the New Mexico Constitution, has meant that for at least the last one hundred and fifty years, the only claimants to water in rivers and streams were persons who diverted water and put it to beneficial use to earn a livelihood. That circumstance has changed dramatically over the past few decades in that a large segment of society considers that rivers and streams, as mobile parts of the ecological systems that surround them, have independent value unrelated to the value of water diverted and consumed from the streams by beneficial users. The greatest manifestation of this perspective is the passage of the Endangered Species Act (ESA), 16 U.S.C. § 1531 *et seq*, that lends its protection to fish species whose existence is being jeopardized by contemporary acts of humankind. The acts of man that jeopardize fish species have been undertaken since development of dam building technology at the beginning of the last century, and although not intended to jeopardize species, have often had that result. In the middle Rio Grande, efforts by environmental groups (including Defenders of Wildlife, Forest Guardians, New Mexico Audubon Council, Sierra Club and Southwest Environmental Center (Environmental Groups)) and the United States to protect one species—the Rio Grande silvery minnow (*Hybognathus amarus*) (minnow)—created a political and legal firestorm that lasted a decade. That epic struggle between irrigation water users and proponents of the species is often called the "Minnow Wars."

Rivers of the Western United States historically have delivered water on the basis of feast or famine—floods or droughts. The mountain snowpacks from above 9,000 feet elevation constitute frozen upstream reservoirs that release the winter's bounty of precipitation as temperatures rise. The rate at which temperatures rise, the amount of wind, and a host of other factors determine both the timing and the quantities of water that are released downstream. A raging torrent in April can be reduced to a trickle or only subflow of a river bed by late July. These variable streamflows have had multiple effects on stream morphology in alluvial basins. These include delivery of large quantities of silt from upstream, wide-braided stream systems with shallow waters warmed by the sun, overbank flooding in wide ecosystems that sustain willows, cottonwoods, and bird species, and provide opportunities for minnow species to spawn.

The hydrograph created by snowmelt stream systems with extreme variations in flow over hundreds of years resulted in the evolution of species of fish adapted to these flow variations. These species have adapted to and prefer for survival slow-moving, shallow, braided stream systems with overbank flooding and high flow periods of cold snowmelt in the spring that promote spawning. They also are capable of surviving in circumstances where there is river drying, when some fish are captured in isolated pools. The species have adapted to follow drying systems upstream to sources of more or less permanent water supply.

While species have adapted and even prospered in highly variable streams, water supply demand for humans lacks this flexibility. Water that has run past a municipality or farm in April is of no value to that municipality or farm in July. Reservoirs have been the answer to smoothing out the hydrograph of available supply. Reservoirs can capture flows and release them at a rate that is consistent on average with agricultural and municipal demand. Although evening out the high flows and low flows in a stream hydrograph may make for reliable municipal and agricultural supply, it also directly affects the viability of species that have relied on the historic variation.

In New Mexico, for example, in the main stem of the middle Rio Grande between Cochiti Reservoir and Elephant Butte Reservoir, the hydrograph has changed, in part, because of the construction of El Vado Reservoir on the Chama, a tributary of the Rio Grande, and construction of the Middle Rio Grande Conservancy District (MRGCD) diversion dams and levees on each side of the river in the late 1920s. Major modifications of the natural hydrograph included channelization of the river, increase in river speed and depth, invasion by non-native species, and changes in water temperature. While the river once flooded miles of riverbank in high flows, this no longer occurs. The most dramatic recent change came about in the mid 1970s when Cochiti Dam was constructed for flood control purposes and to help movement of flows downstream by removal of silt. Cochiti Dam aggravated the changes in quality of water (silt load) and the shape of the river bed in the middle Rio Grande by altering rates of flow. What was once a flowing, meandering, shallow river became a much more canalized one with less acceptable habitat for native species.

**The Minnow Wars Litigation**

ROUND ONE: *MRGCD v. Norton*, 294 F.3d 1220 (10th Cir. 2002). The MRGCD Forces Preparation of An Environmental Impact Statement that Considers the Needs of Irrigators and Raises the Inadequacy of the Science Utilized in Determining the Amount of Water Required by the Minnow.

In the mid 1990s, the United States Fish and Wildlife Service (USFWS) expressed great concern for a fish species known by the common name Rio Grande silvery minnow (minnow). It is a stout silvery fish with emerald reflections measuring up to 3½ inches. It was once common in several western rivers, including the Rio Grande and the Pecos. At the time of its listing as endangered in 1994, the minnow was confined to a short stretch of the middle Rio Grande representing five percent of its historic range. As part of the panoply of protections available to listed species, the Secretary of Interior is obligated to designate what is considered "critical habitat" essential to the conservation of the species. The designation of the critical habitat for the minnow was delayed until July 1999 due in part to the absence of funding for the USFWS. When it was finally completed, the critical habitat designation consisted of 163 miles of the main stem of the middle Rio Grande—cutting directly through the heart of the MRGCD where virtually all of the agricultural diversions took place from four diversion dams. Included in the designation was the finding that approximately 70 percent of the minnow population lived below the southernmost MRGCD diversion point at San Acacia Dam. It became obvious to all concerned that the irrigators in this reach were destined to face direct competition for water from the minnow.

Once the critical habitat designation was made public, both Environmental Groups and the MRGCD expressed great concern regarding the substance of the designation. The Environmental Groups considered the habitat to be under-inclusive. They became more vocal and confident in their legal positions because litigation was being successfully pursued against other irrigation districts around the country including in California, most notably near Klamath Falls, Oregon, where a federal court ordered that waters be left in river systems to sustain fish species rather than be delivered to irrigation headgates. Although those cases were lodged in the United States District Courts within the Ninth Circuit Court of Appeals, plainly they were viewed by Environmental Groups in New Mexico as a part of a national trend and it was considered that they would be followed in the District Courts of the Tenth Circuit Court of Appeals—the courts having federal jurisdiction in New Mexico.

Because the designation of critical habitat was a federal action affecting the environment, both the MRGCD and the Environmental Groups were of the view that a full evaluation of the environmental impacts (EIS) should be made under the National Environmental Policy Act, 42 U.S.C. § 4321 *et seq.* (NEPA) The Environmental Groups considered that the effects on the species should be fully explored, while the MRGCD, on behalf of its irrigators, believed that the possible negative effects on irrigators should be fully explored. The USFWS did not share this view and issued only an Environmental Assessment (EA). The MRGCD wrote the USFWS and asserted that an EA was wholly inadequate because it failed to consider the economic effects on any possible curtailment of use of water by MRGCD irrigators. The Environmental Groups did the same, asserting their concerns. The state and others also expressed concern. Ultimately, the MRGCD filed suit requesting an order that a formal EIS be prepared and others joined the litigation.

The United States District Court ruled in favor of the MRGCD and others and ordered the United States to prepare an EIS. The matter was appealed to the Tenth Circuit Court of Appeals, which affirmed the trial court. The victory given the MRGCD by Senior District Court Judge Edwin Mechem, fully accepted the MRGCD's theory that the definition of the environment affected by a critical habitat designation included more than just the protected species. In affirming the case, the Tenth Circuit Court of Appeals concluded that an EIS was required, *inter alia*, because: ". . . the context of the designation is such that its effects will be felt locally in the Middle Rio Grande valley. (Citations omitted). Given the aesthetic, economic, ecological and cultural value of agriculture to the region, even a loss of 2,000 acres of irrigated farmland is significant."

At the urging of the MRGCD, the Tenth Circuit also took note of the lack of clarity in the evidence presented of the actual quantities of water required by the minnow to survive within the habitat. How much the minnow required was critical, because the middle Rio Grande is a fully appropriated stream. For every drop taken for the minnow, a reduction in water for irrigators was inevitable. The Court noted, "The wide disparity in the estimates of water required for the designation and the associated loss of farmland acreage, indicate that a substantial dispute exists as to the effect of the designation."

Requiring a) the development of a full-blown EIS that considered the impact on irrigators, and b) the judicial acknowledgement of the discrepancies as to the knowledge base related to the minnow's actual water needs were two major victories for the MRGCD. This decision forced the USFWS to consider the role of irrigators in the calculus of preparing a designation of critical habitat and it judicially acknowledged the MRGCD's position that as of the date of the EIS, there was no clear scientific determination of the flows needed by the minnow to avoid jeopardy.

The efforts of MRGCD legal counsel in litigating this case were recognized by the United States District Court when it awarded the MRGCD attorneys fees against the United States in the amount of $191,000 said to be the largest amount awarded against the USFWS as of that date.

> "counsel for the MRGCD [Law and Resource Planning Associates, PC] needed an understanding beyond environmental law and water law. Counsel needed, in addition to several areas of the law, to be conversant in, if not hold a thorough understanding of stream morphology, river management, biology, history, other endangered species in the area, Indian water rights and interstate compacts committing the distribution of Rio Grande water. Thus, in order to direct this case toward a successful end, more than one well-experienced counsel of considerable skill was required. . . . The law involved in the case included developing aspects of environmental, endangered species and water law, as well as administrative and constitutional law; and it was critical at all stages of the case to be well-versed in the most recent cases in the field. Plaintiffs' [MRGCD] briefs and letters indicated an impressive professionalism and competence in each of the fields of law required to pursue their interests, and the briefs from a legal, as well as scientific perspective were well-done in all regards."

> . . . "Without question, Plaintiff MRGCD has prevailed on the

merits. I consider MRGCD as a prevailing party not only on ESA claims, but also on causes of action grounded in NEPA and the APA [Administrative Procedure Act]. Additionally, MRGCD has contributed substantially to judicial review and to the implementation of the Endangered Species Act in accordance with its goals; and MRGCD is clearly entitled to attorneys' fees and costs pursuant both to the ESA and the EAJA [Equal Access to Justice Act]. With regard to the latter [the EAJA], MRGCD has succeeded on the merits (a) against the government, (b) when the government's position was neither reasonable nor justified. The Memorandum Opinion and Order which terminated this litigation expressly concludes that Defendants' actions were arbitrary and capricious and that the position taken by Defendants which precipitated the case was not substantially justified."

Mem. Op and Order Awarding Attorneys' Fees, 6-7, *Middle Rio Grande Conservancy District v. Norton*, No. CIV 99-870, 99-872, and 99-1445 M/RLP (Cons.) (D.N.M. June 18, 2001).

ROUND TWO: *Minnow v. Keys I*—The Environmental Groups Sue to Give the USBOR Discretion to Take MRGCD Water for the Minnow. They Prevail in the Trial Court, but the Decision is Vacated and Congress Comes to the Rescue of the Irrigators. *Minnow v. Keys*, 355 F.3d 1215 (10th Cir. 2004).

Not long after the ruling in *MRGCD v. Norton* and the issuance of a revised final EIS that included within it the effects on the irrigators, the issue began to heat up further. The pressure from Environmental Groups for protection of the minnow intensified and statements were made by Environmental Groups and "experts" that unless the Rio Grande was kept wet from Cochiti Dam to Elephant Butte Reservoir, the species would be extirpated entirely within the middle Rio Grande. Because of a reduced snowpack from the winter before, the river began to dry below San Acacia Dam. The common wisdom among the Environmental Groups and within some segments of the biology community was that because there would be low runoff during the irrigation season, and because farmers were planning to irrigate from these low flows, the direct enemies of the minnow were the irrigators of the MRGCD. Environmental Groups believed that the United States Bureau of Reclamation (USBOR) had the absolute power to control diversions. Continued diversions by irrigators would cause drying of the river. The environmental experts concluded that any drying below San Acacia Dam would make the minnow extinct. Therefore, the MRGCD diversions must be stopped by the USBOR. Better a few farmers lose their crops than lose the minnow forever.

The MRGCD irrigators did not agree that the USBOR had the authority to shut down their headgates, nor did they agree that some drying would cause extinction. The MRGCD retained expert fish biologists that pointed out that the river had been dry below San Acacia Dam for multiple consecutive years in the 1950s and the minnow had survived. Therefore, it was counter-intuitive to argue that if the Rio Grande dried in the current year, the minnow would become extinct. Even so, the Environmental Groups' biologists were steadfast in their view that unless diversions were stopped, the minnow would become extinct.

The politics within the USBOR and the USFWS became even more heated as local USBOR officials and USFWS officials faced pressure to "save" the minnow. Matters were brought to a head when the USBOR concluded that under the ESA, it was obligated to determine whether its actions would affect the minnow and possibly place it in jeopardy. If the USBOR concluded its action would cause jeopardy, then the USBOR would be obligated to consult with the USFWS and take action to ensure that its federal actions did not cause injury to the listed species. The federal action being reviewed was whether to continue to allow MRGCD irrigators to divert water for their crops.

The procedure for evaluating federal agency effects on species and the consultation process under the ESA is complex. When a species such as the minnow is listed as endangered or threatened, the listing triggers, *inter alia*, the provisions of Section 7(a)(2) of the ESA. This section requires that any responsible federal agency taking action that could affect the species, in this case, the USBOR, must evaluate the effects of its actions in relation to the listed species. The outcome of the USBOR evaluation is a determination by the federal agency whether it is a) engaging in *discretionary* operations affecting the species and b) whether the exercise of this discretionary authority to carry out its operations is likely to jeopardize the existence of the species. The written product of this evaluation is a biological assessment (BA) of the effects of the federal agency's actions on the species. The BA contains both legal and factual conclusions. The legal conclusions answer the question of whether the actions of the USBOR being evaluated are discretionary ones. If the USBOR determines that it has the legal discretion to change the activities it is currently carrying out and if the discretionary change could lessen or alter the effects on the listed species, then the ESA requires an evaluation of the extent of the USBOR's effects on the species to be conducted by the USFWS. The document for measuring and reporting whether the USBOR's actions place the species in jeopardy is called the Biological Opinion (BiOp).

Because the BA prepared by the USBOR reached the conclusion that it had authority to exercise discretion in regulating MRGCD irrigators' diversions from the river and that diversions from the river affected the species, the USBOR and the United States Army Corp of Engineers (the Corps) were obligated to consult with the USFWS.

The outcome of the consultation among the three agencies was the BiOp. In the simplest terms, the minnow BiOp was designed to determine formally whether ongoing or proposed federal actions of the federal agencies were likely to jeopardize the existence of the listed species. The BiOP is a complex technical document prepared exclusively by the USFWS. It cannot contain the subjective views of the agency; rather it must be based upon the best scientific and commercial data available. If jeopardy might be caused by federal actions, under the ESA, the federal agency must modify its proposed or ongoing actions. In short, the ongoing federal actions jeopardizing the species must be stopped unless the parties determine that there is a reasonable and prudent alternative to stopping the federal actions. The reasonable and prudent alternative is designed to accommodate the needs of the federal agency and its constituents while not causing the "take" (death) of an inordinate number of the species. A reasonable and prudent alternative is one that allows federal action while causing what the BiOp determines to be an acceptable "incidental" take of the species. While, normally, the take, or killing, of an endangered fish would be a crime, under the ESA, the BO can preclude prosecution of persons who cause no more than the incidental take allowed by the BiOp. This is called providing take coverage.

The MRGCD was stunned by the substance of the BA provided by the USBOR. For the first time in over thirty years, the USBOR asserted that it owned all of the diversion dams of the MRGCD and furthermore, because it considered the MRGCD a "federal project," the USBOR had absolute discretion to determine whether irrigators should receive water from the river or whether water should simply be left in the river for the species and the irrigators' headgates closed.

As discussed below, the MRGCD argued loudly and forcefully that the USBOR had no discretion to take water from irrigators, because the MRGCD was not a federal project, delivered no native federal water to irrigators, did not own the MRGCD works and that the act of diverting water to irrigators was an exercise of irrigator-owned water rights under New Mexico state law. Not only did the MRGCD disagree with the USBOR on the law, it disagreed on the facts. The MRGCD argued that there was no sound scientific basis for requiring that the Rio Grande remain wet from Cochiti Dam to Elephant Butte, as suggested by the BA, USFWS and environmental group biologists, and by other recent studies of the minnow by contract biologists at the University of New Mexico. To ensure there was no federal financial connection to the project, the MRGCD paid off the balances on two outstanding federal contracts that remained with the USBOR—one contract in 1951 for rehabilitation of MRGCD works and a second amendatory contract in the 1970s for payment of a portion of the cost of the San Juan-Chama project. The MRGCD paid off the 1951 contract immediately upon hearing the USBOR claimed ownership of MRGCD works. The MRGCD also attempted to pay off the balance on the San Juan-Chama contract, but the USBOR refused to take the check and accept early payment. An act of Congress introduced by Senators Domenici and Bingaman forced the USBOR to accept the MRGCD's payment for the San Juan-Chama contract, thus completely eliminating any financial dependence of the MRGCD on the USBOR.

As required by the federal law, the BiOp was issued by the USFWS. It reached essentially the same conclusions as the BA, finding that the USBOR had discretion to take water from irrigators and provide that water to the species. It further suggested that this may be required under certain conditions. However, it did not require that the river be kept wet in the middle Rio Grande during the entire irrigation season but allowed for an exercise of choices in the USFWS on this issue.

Unsatisfied with the aggressiveness of the USFWS, USBOR and the Corps, the Environmental Groups filed suit against these three federal agencies arguing essentially that the agencies violated multiple provisions of the ESA by not taking control of MRGCD and ceasing irrigation diversions so that the river remained wet from Cochiti Dam to Elephant Butte. Recognizing that shutting down irrigation could destroy thousands of acres of crops and cost them their livelihoods in many cases, the MRGCD irrigators were incensed. The MRGCD intervened as a defendant in the litigation arguing that the USBOR had no discretion to take the private water of irrigators, that the United States did not own the MRGCD works, and finally, that the assertions of the Environmental

Groups requiring that the river remain wet for its entire reach were not supported by the best scientific and commercial data available as required by the ESA.

The anger of the irrigators was palpable at every public meeting. In response, the Environmental Groups vilified the irrigators as wasteful enemies of endangered species. Irrigators indicated that they would fight any attempt by the United States Marshals to dry up their fields. Violence almost occurred when, on July 6, 2000, Mr. Subhas Shah, Chief Engineer for the MRGCD, was ordered by the USBOR to close the irrigation headgate at San Acacia Dam. Irrigators got wind of the order and broke the lock off the gate and defied the edict of the USBOR. At the diversion dam, Mr. Shah climbed into the bed of a pickup and assured the irrigators that he would not allow the United States to take the water, and that the MRGCD would fight the issue in court. The ditch remained open while matters cooled down.

Months of settlement negotiations took place through a federal magistrate pursuant to an order of the federal court as part of the federal litigation. As a result of the mediation, the USBOR was able to purchase and the MRGCD was able to borrow San Juan-Chama water from the City of Albuquerque and others that was released into the river. The water released eased the effects of the drought on the irrigators and provided water to the minnow. This relieved the pressure to take water from MRGCD irrigators but it did not resolve the issue. While the MRGCD cooperated with the federal agencies, it continued to refuse to close any headgates.

Ultimately, and after a great deal of procedural maneuvering by the parties and various interim rulings, federal judge James Parker ruled, in effect, that the USBOR had discretion to divert water away from irrigators for use by the endangered species such as the minnow. Judge Parker also made findings as to the needs for water supportive of the claims of the Environmental Groups that if there were river drying, the minnow would become extinct. He also found that water purchased by the City of Albuquerque under the San Juan-Chama Diversion Project was available for the minnow. The federal judge ordered that water must be released from upstream reservoirs rather than stored for irrigators. The water that the Court ordered released included not only native Rio Grande water, but also water that had been diverted from the San Juan River Basin as part of the San Juan-Chama Project. The ruling awarding the USBOR discretion over MRGCD water was appealed. The Tenth Circuit Court of Appeals initially, by a two to one vote, affirmed Judge Parker's decision. However the MRGCD, the United States, and others moved for a re-hearing before the entire panel of justices of the Tenth Circuit Court of Appeals. Before the matter could be reviewed by the entire panel, the matter was vacated as moot by the judges that had initially ruled in favor of the Environmental Groups.

The matter had become moot on appeal because in May of 2003, the USFWS issued a new 10-year BiOp that avoided the discretion issue, by allowing for an evaluation of total river depletions from all sources independent of sources of water. To produce this BiOp, all interested agencies—the USFWS, USBOR, and the Corps—participated in the consultation, and agreed on reasonable and prudent alternatives that would not take irrigators' water. The BiOp also set permissible levels of incidental take of the species that would not violate the ESA.

Shortly thereafter, the New Mexico congressional delegation began to assist. In December of 2003, at the request of the City of Albuquerque and other owners of San Juan-Chama imported water, Congress passed a rider to the Energy and Water Development Appropriations act. The so-called "Minnow Rider" denied USBOR the right to take imported water to meet minnow needs. The Minnow Rider also contained a Congressional determination that compliance with the 2003 BiOp by irrigators and others for a two-year period met all of the requirements of the ESA. A series of additional legislative actions in 2005 extended the Congressional conclusion that compliance with the 2003 BiOp constituted compliance with the ESA so long as the 2003 BiOp was in force.

The opinion of the Tenth Circuit Court of Appeals reversing and vacating the opinion that had previously affirmed the USBOR's discretion to take MRGCD irrigators' water was the second significant judicial victory for the MRGCD. The appellate decision meant that the Environmental Groups continued to have no final judicial ruling that could compel the MRGCD to cease irrigation in favor of the minnow. This judicial victory coupled with the Congressional action set the stage for the next legal fight. The 2003 BiOp mandated action that the Environmental Groups and the federal district court had previously argued vehemently would result in the extinction of the minnow. A new federal lawsuit from the Environmental Groups attacking the 2003 BiOp was expected at any moment.

ROUND THREE: *Minnow v. Keys III*—The Environmental Groups Do Not Sue, the USBOR Changes its Position on its Ability to Take MRGCD Irrigation Water, and the Tenth Circuit Court of Appeals Rules that All Previous Rulings of Judge Parker Adverse to the MRGCD on the Discretion Issue are Void and that the Environmental Groups' Complaint Must be Dismissed. *Minnow v. Bureau of Reclamation*, 601 F.3d 1096 (10th Cir. 2010).

Surprisingly, even though they had previously argued that drying of the river, as allowed by the 2003 BiOp, would be Armageddon for the minnow, the Environmental Groups did not challenge the 2003 BO. Accordingly, the MRGCD, the United States, and others moved to have all previous decisions entered by Judge Parker on the discretion issue vacated entirely and removed as precedent for any future case. The rationale was that there was no longer a live "case" wherein the discretion issue needed to be decided. Without a live case, the Court had no jurisdiction. The Environmental Groups opposed the Motion and argued that the discretion rulings of Judge Parker should be maintained as precedent and binding against the MRGCD because the issue might arise again in the future. Once again Judge Parker ruled in favor of the Environmental Groups and against the MRGCD and others, concluding that his previous decision should remain as precedent. Once again the MRGCD, the United States, and others appealed.

By this time, the Tenth Circuit Court of Appeals appears to have had enough. All of Judge Parker's decisions in favor of the Environmental Groups were vacated and could not be cited as precedent. Although explaining that in addressing all of these complex issues, Judge Parker had done his very best to deal with a complex problem, the Court of Appeals reversed his refusal to vacate his opinions against the MRGCD on the discretion issue. The appeal raised some novel issues that reflected how personal the issues had become, not only for the litigants, but also for the Judge. One of the more unusual reasons asserted by Judge Parker as to why his decision should remain as precedent was that the decision was necessary to rebut the comments made by the Mayor of Albuquerque in the *Albuquerque Journal* newspaper. The Mayor had stated in the newspaper that Judge Parker's decisions amounted to stealing San Juan-Chama water from the City. Judge Parker considered what amounted to an exercise of his first amendment right by the Mayor as so egregious that it justified leaving his decision in place in an otherwise moot case.

> "[Mayor Chavez's allegations that 'someone's stealing our water'] could easily be interpreted by the public as an accusation that criminal intent underlay the ruling, and granting the vacatur motions may be construed by some as validation of that position.

Accusing a judge of committing the crime of theft in making a judicial ruling is, of course, a very serious matter." Memorandum and Opinion, fn. 9 (November 22, 2005).

The Tenth Circuit Court of Appeals was not persuaded.

A second unique factor in the appellate case was that the USBOR reversed the position it had taken earlier on the discretion issue in the BA. In the BA, it had concluded that the USBOR had discretion to take water from MRGCD irrigators. In this latest minnow case, the USBOR took the opposite view. *See* Federal Defendants' Brief in Chief in *Rio Grande Silvery Minnow et al. v. Bureau of Reclamation, et al.*, Tenth Cir. Ct. App., Nos. 05-2399, 06-2020, 06-2021 (Consolidated), pg 11 (June 2006). ("[T]he United States never acquired both of MRGCD's state-permitted rights to use and store native Rio Grande water.") This is true regardless of any ultimate outcome concerning ownership of the physical works for water deliveries as the Bureau has also conceded in court filings. *See also* Federal Defendants' Reply Brief, Point IIIA, pg. 19. ("The United States' Ownership of the Rio Grande Project *does not expand discretion* under the contracts at Issue.")

In reversing Judge Parker this final time, the Tenth Circuit was somewhat indelicate in its language: "We need not decide whether *any one of the district court's manifest errors of judgment* discussed above would, standing alone, constitute grounds for reversal of its order denying vacatur." *Rio Grande Silvery Minnow v. Bureau of Reclamation*, 601 F.3d 1096, fn. 22 (10th Cir. 2010). The Court went on to conclude that the collection of manifest errors of judgment required reversal. Significantly, the mandate from the Tenth Circuit to the district court was to not only vacate all previous opinions on the discretion issue but to dismiss all counts of the Environmental Groups' Third Amended Complaint, thus ending the case without a drop of water ever being involuntarily taken from irrigators for use by the minnow.

**Lessons Learned From the Minnow Wars: Oversimplified Assumptions and Politics Can Be Major Drivers in Environmental Litigation.**

The underlying dispute that created ten years of litigation between Environmental Groups and consumptive users of water went beyond the simple question of whether the minnow could be saved from extinction. It raised latent conflicts between sincerely felt fundamental value judgments as to the best use of the Rio Grande. Proponents on both sides of the issue were convinced they were supporting the correct policy for the highest and best use of the river. The Environmental Groups were absolutely convinced of the truth of two sets of facts. The first was that irrigators in the middle Rio Grande valley were wasting water. The argument was made that the quantity of water consumed by the crops of irrigators was the equivalent of covering each acre with ten feet of water. This inaccurate figure for measuring the consumptive use of water by MRGCD irrigators was generated by summing the diversions of all four MRGCD diversion dams. The correct facts were that after water was diverted to fields and consumed through evaporation by crops, the excess water soaked into the ground, and much of it returned to the river though drains or through the soil. For example, the New Mexico State Engineer calculates that almost one-third of the water diverted onto fields in the MRGCD returns to the river as "return flows." Therefore, each diversion below the previous one diverted water that had previously been diverted but had returned to the river as return flow.

Thus, the diversions could not be summed, but rather it was the sum of the actual amounts consumed in each reach that was the accurate number for depletions from the Rio Grande. It is likely that the confusion from this double and triple accounting affected Judge Parker's ruling because it was included in a report submitted to him and referenced in his decisions. A subsequent report describing the MRGCD's accounting methods was excluded by the Court as not being timely submitted as part of the administrative record. The MRGCD report was not submitted earlier because the MRGCD was not allowed to participate in the formulation of the BA or the BiOp. The second fact that the Environmental Groups firmly believed was based on the views of their experts. They believed that if the river were allowed to dry at any point below San Acacia Dam, the minnow would become extinct.

In contrast, the middle Rio Grande irrigators were of the view that their diversions needed to stay exactly as they had been in the past and that there was no room for compromise. Their fears were based in part on stories of massive financial losses in the Klamath, Oregon basin where irrigators were driven out of business when water was taken from them for endangered species. Indeed there was the view among some that irrigators could never find common ground with either the Environmental Groups or the USFWS. Just as some Environmental Groups believed that irrigators were indifferent to the needs of the minnow and could care less if it became extinct, some irrigators were convinced that under the ESA, the Environmental Groups, the USFWS, and the USBOR were all of the view that saving irrigators' livelihoods was essentially irrelevant.

As is often the case with such polarized positions, neither was correct. Although the Environmental Groups grossly overstated the facts as to the inefficiency of the MRGCD, they were correct in asserting that it was feasible, through coordinated management of diversions, to support the needs of the minnow while allowing irrigation to continue. Likewise, even though the Environmental Groups were incorrect in arguing that the species would become extinct if the river did not remain continuously wet, their efforts required the USFWS to further evaluate the needs of the minnow and devise systems to minimize "take" of the species when the Rio Grande did recede.

Indeed, valuable research is continuing. And ultimately, the facts that the Environmental Groups did not challenge the 2003 BiOp, that Congress passed the "Minnow Rider" providing irrigators congressional protection for their diversions, and that in the final minnow appeal, the USBOR and USFWS modified their positions on the discretion issue to preclude federal agencies from taking irrigators' water, all demonstrate that rational approaches to these kinds of issues can prevail. There were no white hats or black hats in this struggle. Although they were ultimately not successful, the Environmental Groups were represented by very able and sincere legal counsel that presented their arguments well and persuasively. They had achieved greater success elsewhere in the Ninth Circuit Court of Appeals than they did in the middle Rio Grande. The Tenth Circuit Court of Appeals was not as receptive to their arguments. They and their clients were acting in reliance on the advice of their experts that if they did not file suit, the minnow would become extinct. But in the end, their experts were wrong and the irrigators of the middle Rio Grande valley were correct in not sacrificing their water and water rights.

The Minnow Wars demonstrate in some ways, in gross relief, a major flaw in the ESA—a species can only be accorded full mandatory protection when it is nearing extinction. But if the USFWS can only take action when extinction is imminent, there is no time to develop the science required to know how to avoid extinction. Once listing occurs, the ESA requires immediate and draconian action to protect it. But, of course, the fact that the USFWS knows that it is legally obligated to protect a species to avoid extinction does not mean that it knows what action to take.

Because the minnow is a fish species, the only variable easily modified to change the minnow's circumstance is the quantity of water in the river. Just because the quantity of water in the river is the only variable at hand, does not mean that changing it will increase the survivability of the species. In the case of the minnow, the simplified correlation between the facts that the minnow had lost population, and that the river has been drying in low population years do not inevitably lead to the conclusion reached by the biologists that the Rio Grande must be kept wet through its entire reach to avoid extinction. But given the ESA's call for action, the USFWS, based on the mandates of Judge Parker, had little choice but to increase the quantities of water. According to Judge Parker, to not add water to the river was a certain recipe for extinction.

A second value judgment was also at work. There were many who believed that a flowing river all summer would provide overall environmental benefits even if it did not make the species better off. Thus, there was political momentum to ensure an instream flow in the middle Rio Grande by restricting diversions for agriculture. The minnow was viewed as the lever for freeing up this instream flow. The sum of these two motivations appears to have caused biologists for the Environmental Groups to make predictions for extinction of the minnow far beyond their scientific understanding. The initial testimony of the environmental experts in the case offered an all new meaning to the phrase "political science."

For example, the first "expert" to address the water needs of the minnow on behalf of the Environmental Groups was Dr. Dean Hendrickson. He filed a declaration under oath swearing that to avoid extinction of the minnow...

> "[A]t a minimum, the Rio Grande needs to retain continuous flow throughout the reach from Cochiti Reservoir down to Elephant Butte Reservoir. *This means that the streambed must remain wetted, in at least one flowing channel throughout the length of the river.* Allowing flows to recede, with consequent formation of isolated pools and dry riverbed must be avoided." (J.A. Vol. IV at 647-648).

This was consistent with the allegations of the Environmental Groups in their Complaint that

> "[u]nder the June 29[th] BO/RPA, river drying will commence shortly and continue much of the remaining summer 2001 as well as in summers of 2002 and 2003. *This is a certain prescription for extinction of the silvery minnow*..." (J.A. Vol III at 422-23).

Another expert for the Environmental Groups, John S. Pittenger, testified on September 2, 2002 that

> "[o]nset of intermittency *and drying in the Angostura and Isleta reaches this year, in addition to the dewatering of the San Acacia Reach would likely result in extirpation of the Rio Grande Silvery Minnow in the Middle Rio Grande.*" (J.A. Vol IV at 701).

Judge Parker accepted all of these conclusions, holding in his September 23, 2002 Opinion that

> "The present biological condition of the silvery minnow is so severe that *extensive drying in the San Acacia Reach could result in the extinction of the silvery minnow in the wild.*" (J.A. Vol II at 221).

Ironically, after all of the litigation in reliance on these scientific "facts," the USFWS in the 2003 BiOp concluded that river drying below San Acacia through staged recession and species recovery would not lead to extinction. This conclusion was not challenged by the experts or the Environmental Groups. Upon final approval of the 2003 BiOp, the Environmental Groups' experts and the groups themselves remained silent. The minnow is not extinct even though river drying has occurred every year since 2003 and the minnow population is no worse off.

On the institutional front, matters have improved dramatically. Pursuant to federal legislation requiring a collaborative approach to this problem, the USFWS, USBOR, the MRGCD, environmental representatives, the New Mexico Interstate Stream Commission, and irrigator representatives have been collaborating on a plan to further protect the species. The plan is to replace the 2003 BiOp with a better and more inclusive BiOp that will integrate principles directed toward recovery of the species. There has been extensive discussion of utilization of adaptive management to improve systematically the prospects for the minnow. Finally, these entities are developing models utilizing Population Viability Analysis to determine the variables that when modified, can increase the probability of ensuring viability of the species. These variables may include rates of water flow, but they also include timing of releases and instream bank morphology modification, to name a few others, instead of simply maintaining a continuously flowing river. At least as of now, without the intervention of more "political" science, the future of the minnow and the irrigators is looking up.

This has been a long and difficult path, but the end result is a much better circumstance for species and irrigators alike. While one would have hoped for rational discourse from day one, it is far from clear given the uncertainties and the stakes in the outcome that a different path could have been found. Justice Holmes writing for the Tenth Circuit Court of Appeals, while not perhaps understanding the poetic significance of his language in explaining the inevitability of the conflict, observed:

> "This case involves one battle in a prolonged war over a finite and elemental resource—Rio Grande water. The needs of the plants and animals that depend upon this water for survival are in tension with the needs of the human inhabitants of the Middle Rio Grande valley ("the Valley") who depend upon the water for daily living and commercial and agricultural activities." Judge Holmes' Slip Opinion at 4.

Justice Holmes was correct. The Endangered Species Act is not a law that calls for species protection without regard to the needs of uses by others. A practical balance is required for it to work. Persons who used water before passage of the Act cannot behave as though Congress has not spoken on the issue of species protection. It has. But Environmental Groups cannot presume that the baseline is a river with no human habitation. In the case of the minnow, the "war" has now moved out of the courtroom and around a collaborative negotiation table. The outcome of this collaborative effort will turn on adherence to the most critical factor enumerated by Congress in passing the Endangered Species Act—that the actions taken by all shall be based upon the "best scientific and commercial data available." If the minnow's future is shaped by strict adherence to this scientific standard, then its future is bright. However, if any party seeks to substitute "political" science for the best science, or if any party reverts to a previous polarized position, then the matter will likely wind up again in court and both the minnow and the irrigators will be the worse for it.

# 9

# Struggle Over Pueblo Water Rights: The Aamodt Case

## John W. Utton

**Introduction**

The *Aamodt* case holds the unenviable distinction of being the longest pending case in the federal court system, taking up over 40 feet of shelf space in the federal courthouse. Filed in 1966, *State ex rel. State Engineer v. Aamodt* is the water rights adjudication of the Pojoaque Basin north of Santa Fe, including the Rio Pojoaque and its two major tributaries, the Rio Tesuque and Rio Nambe. Despite active litigation for three decades, the case was still far from resolution in the year 2000 when the parties sought to stay further litigation in favor of settlement negotiations. Under the settlement reached in 2006 and passed by Congress in 2010, the claims of the four Pueblos of Nambe, Pojoaque, San Ildefonso, and Tesuque will soon be resolved by the federal court's approval of the settlement and entry of a final decree.

Despite *Aamodt*'s age, the sources of legal conflict are much older, arising well before statehood. First with respect to land ownership and later water rights, Pueblo Indian property rights in New Mexico did not fit the mold for determination of tribal rights adopted elsewhere in the United States. Unlike other Indian tribes, the New Mexico Indian Pueblos were not re-located to federal reservations created by peace treaty. Instead, Pueblo Indians continued to inhabit traditional villages along the Rio Grande Valley. As successor sovereign, the United States confirmed lands to the Pueblos claimed to be grants from the King of Spain and recognized by the Republic of Mexico.

This different history of Pueblo land status caused the U.S. Supreme Court in 1876 to hold that Pueblo land was not protected by federal law prohibiting the alienation of Indian lands. Before the Court reversed itself 37 years later, non-Pueblo settlers acquired numerous tracts of land within Pueblo grants. Even after Congress' two attempts to redress the fallout from the many adverse land claims within Pueblo grant lands, the status of Pueblo water rights remained unresolved. Again, the fact that Pueblo grant lands were established prior to U.S. sovereignty and without a treaty became the key issue in the *Aamodt* case, as the adjudication court and parties grappled with the nature of Pueblo water rights in the absence of a federal reservation. Elsewhere in the western states, the developing Federal Reserved Water Rights Doctrine or *Winters Doctrine* defined the quantity and characteristics of Indian water rights. In *Aamodt*, the court would chart its own course.

**Early Historical Setting**

The Pueblo Indians of New Mexico have lived along the Rio Grande and its tributaries, using its waters for domestic and irrigation purposes, since long before arrival of European settlers.[1] Historians have opined that Pueblo Indians are the descendants of the Anasazi Indians, who occupied the Four Corners region of what is now New Mexico, Arizona, Utah, and Colorado. The Anasazi abandoned the Four Corners area around 1200 A.D., apparently due to drought, and eventually settled in the Rio Grande Valley. The Pueblo villages along the Rio Grande and its tributaries first were settled in the early to mid 1300s.[2] Zuni Pueblo is believed to have been inhabited beginning between 900 to 1,000 A.D.[3]

When Don Francisco Vasquez de Coronado and the Conquistadors arrived in the Rio Grande Valley in 1540, they found an estimated 25,000 acres of land under cultivation by Pueblo Indians.[4] Pueblo members occupied and cultivated their lands throughout the Spanish colonial period and throughout Mexican rule from 1821 to 1846, and continuing during the sovereignty of the United States from 1846 to the present.[5]

During the Spanish colonial period, Spanish law recognized a legal ownership interest in the Pueblos to their lands and waters, and Spanish law protected this interest.[6] Each Pueblo owned its land as a community and could not alienate its land without a royal license.[7]

Spanish law also recognized and protected a Pueblo's use of water.[8] Use of water was administered under the *repartimiento* system, a quasi-judicial and administrative proceeding in which a government official apportioned water among all water users based on equitable principles, including historical use and need. Under this system, all individuals were accorded water to meet domestic and sanitary needs. Pueblo irrigation rights were based on the principle of *para su susteno*, the quantity of water necessary to grow crops for subsistence. Thus, need was determined from customary and actual use, and the Pueblos' "long use was the dominant consideration in the determination of need."[9]

The overall effect of Spanish law on Pueblo water use was to preserve "a first right of *primacia*, to enough water for their needs, for the irrigation of their lands and maintain the community."[10] This right included excess water for future expansion, based on need.[11]

In 1821, Mexico won its independence from Spain. Although Mexican law was different in significant ways from Spanish law,[12] the Pueblo Indians continued to occupy and cultivate their historic lands.[13] Under Mexican governance, the *repartimiento* remained as the system of water allocation in times of shortage, applying principles of equity to Indians and non-Indians. Mexico continued to recognize that the long history of Pueblo water use was the dominant consideration in determining need.[14]

Mexican sovereignty was short-lived, ending in 1846 when the United States occupied the territory. On May 30, 1848, the United States and Mexico entered into the Treaty of Guadalupe Hidalgo to end the "calamities of war" that existed between the two countries.[15] Among other things, the Treaty established the boundaries between the two countries.[16] Mexico ceded property to the United States under the Treaty, including most of what is now the State of New Mexico. Importantly, the Treaty preserved the rights of Mexicans residing in the ceded territory, including their property rights. Article VIII provided in part:

> In the said territories [ceded by Mexico to the United States], property of every kind . . . shall be inviolably respected. The present owners, the heirs of these and all Mexicans who may hereafter acquire said property by contract, shall enjoy with respect to it guaranties equally ample as if the same belonged to citizens of the United States.[17]

Thus, Section VIII of the Treaty provided for "property of every kind" belonging to Mexicans within the ceded territory to be "inviolably respected," and guaranteed that the Mexican owners, their heirs, and "Mexicans acquiring said property by contract," would enjoy the same protections of property as citizens of the United States.[18]

The Pueblo Indians were Mexican citizens under Mexican rule and were considered "Mexicans" for purposes of the Treaty.[19] Therefore, the Treaty afforded the same protection of property rights to the Pueblos that it afforded to other residents of the ceded territory.[20]

Shortly after signing the Treaty of Guadalupe Hidalgo, the United States Congress enacted legislation to protect the rights of Indians in the New Mexico Territory. The Trade and Intercourse Act of 1834 already prohibited non-Indian settlement on Indian lands and provided that Indian lands could only be conveyed by treaty or constitutional convention.[21] In 1851, Congress extended the provision of the 1834 Act to the New Mexico Territory.[22]

In 1858, Congress enacted the Confirmation Act, confirming the New Mexico Pueblos' land titles.[23] The Act instructed the Commissioner of the Land Office to issue patents to the Pueblos as it did in ordinary cases to individuals. The Act provided ". . . that this confirmation shall only be construed as a relinquishment of all title and claim of the United States to any of said lands, and shall not affect adverse valid rights, should such exist."[24]

**Encroachments on Pueblo Lands and Water**

By the time of Statehood, congressional hearings determined that as many as 3,000 non-Indians claimed lands and use of water within Pueblo lands.[25] Much of the non-Indian encroachment on Pueblo lands after U.S. acquisition of the territory was attributable to court decisions holding that federal laws protecting Indians and prohibiting alienation of their lands (and appurtenant water rights) did not apply to the Pueblos. Chief among those cases was the infamous 1876 Supreme Court decision in *United States v. Joseph*,[26] which opened the door for non-Indians to occupy and acquire Pueblo lands without government consent.[27]

In *Joseph*, a non-Indian settler on Taos Pueblo lands was sued for trespass in violation of the 1851 Trade and Intercourse Act. The United States Supreme Court determined that the Pueblos held title to their lands differently from other Indian tribes. The Court observed that in general the United States held ultimate title to tribal lands, recognizing in the tribes only "passing title with right of use" and no right of transfer "until by treaty or otherwise that right is extinguished."[28] But in the Pueblos' case, the Court concluded that the Pueblos held title to their lands dating back to grants under Spanish rule, that title had been fully recognized by the Mexican government and by the Treaty of Guadalupe at the time of transition to American government, and that reports from the New Mexico surveyor general, made at the direction of Congress, recommended confirmation of title to the Pueblo of Taos. This led the Court to hold that the Pueblos held title in the same manner as an individual would hold perfect title, and that this title was superior to that of the United States, and that the United States had disclaimed the right of present or future interference except as might be exercised against any person holding competent and perfect title.[29]

Decisions of New Mexico territorial courts during this time also had the effect of reducing federal protection of Pueblo lands. In 1869, the territorial court held in *United States v. Lucero* that a trespass action brought by the United States on behalf of Cochiti Pueblo against Lucero under the Nonintercourse Act could not be sustained because the Pueblo Indians were not subject to the Act.[30] In 1900, the territorial court relied in part on *Joseph* to reject a trespass claim on lands of the Pueblo of Nambe.[31] In 1907, the territorial court in *United States v. Mares* held that individuals could not be prosecuted for violating an 1897 federal act prohibiting the sale of intoxicants to Indians because the Pueblos were not "Indian tribes" under the federal acts.[32]

**Congress Intervenes**

Following the judicial branch's role in allowing encroachment on Pueblo lands, Congress took its turn to solve the increasingly difficult problem of land status within Pueblo grant lands. Beginning with the enabling act for New Mexico statehood and then extending to two Pueblo lands acts, Congress achieved significant redress in the early part of the 20th Century, but still left open a number of issues, especially related to water rights.

In considering legislation approving New Mexico statehood, Congress made clear that Pueblo Indians are "Indians," at least within the meaning of the 1897 Act prohibiting the sale of intoxicating liquor on Indian lands. As a condition of statehood, New Mexico's Enabling Act, Section 2, extended coverage of the 1897 Act to Pueblo Lands.[33] The State assented to this provision, effectively overturning the territorial court's previous holding in *United States v. Mares*. Shortly thereafter, the U.S. Supreme Court in 1913 disapproved the *Joseph* decision in *United States v. Sandoval,* holding that the Pueblos were "Indian Tribes" within the meaning of the Commerce Clause and therefore Congress had authority to prohibit intoxicating liquor on Pueblo lands under the New Mexico Enabling Act.[34] The *Sandoval* Court found that Pueblo Indians are

"Indians in race, customs, and domestic government" and that the United States has always treated them as dependent Indian communities.[35] Thus, while *Joseph* concluded that Pueblo title is not subject to government interference, *Sandoval* held that the Pueblos are Indians subject to federal protective powers and that *Joseph* could not be regarded as holding that the Pueblo Indians are beyond the range of congressional power.[36] The applicability of the Nonintercourse Act to the Pueblos was finally put to rest in 1926 when the Supreme Court expressly found that the Pueblos were "Indian Tribes" within the meaning of the Act.[37]

Congressional action to confirm that Pueblo lands are subject to federal protections of Indian lands placed non-Indian settlers in a precarious position. Even good faith settlers faced ejectment lawsuits. Congress responded by enacting the Pueblo Lands Act of 1924 to resolve questions of Pueblo title that arose due to non-Indian encroachment on Pueblo lands and to provide compensation to Pueblos for lost land and appurtenant water rights. "The stated purpose of the Act was to 'settle the complicated questions of title and to secure for the Indians all of the lands to which they are equitably entitled.'"[38] The Act established a Pueblo Lands Board to investigate Pueblo title, determine to what lands Pueblo title had not been extinguished, and to investigate competing claims. The Act required a unanimous Board decision to extinguish Indian title.[39] The board was to file reports describing Pueblo lands for which title had not been extinguished. Based on these reports, the U. S. Attorney General was to bring suit in the United States District Court for the District of New Mexico to quiet title to these Pueblo lands.[40]

Because of claims that the Pueblo Lands Board failed to recommend full compensation to the Pueblos for loss of land and appurtenant water rights, Congress once again took action by passing the 1933 Pueblo Compensation Act. The Act provided additional compensation to the Pueblos for the full market value of the land and appurtenant water right lost as originally directed by the 1924 Act. The 1933 Act also provided that remaining Pueblo water rights could not be lost through forfeiture for non-use or abandonment.

By operation of the Pueblo Land Act of 1924 and the Compensation Act of 1933, the Pueblos lost land and appurtenant water rights to non-Indian settlers but still retained water rights appurtenant to their remaining lands. The exact nature of these water rights would be the focus of stream system adjudications of the Rio Grande tributaries, especially the *Aamodt* adjudication of the Pojoaque Basin.

## The Aamodt Adjudication

Although the 1924 and 1933 Acts resolved disputes concerning land title within Pueblo land grants, the adjudication of Pueblo water rights remained and the stage was set to determine water rights appurtenant to Pueblo lands confirmed under the Pueblo Lands Act of 1924 and based upon the Compensation Act's strong implication that Congress intended to preserve Pueblo water rights for their remaining lands.[41]

At the forefront of the adjudication of Pueblo water rights in New Mexico is the adjudication of the water rights of the Pueblos of Nambe, Pojoaque, San Ildefonso, and Tesuque in the Pojoaque streams system in *New Mexico ex rel. State Engineer v. Aamodt*.[42] The State of New Mexico filed the action in 1966 to adjudicate the waters of the stream system, including the rights of the four Pueblos.[43]

Key questions in adjudicating Pueblo water rights in New Mexico were whether the water rights should be based on state law or federal law, and if the latter, whether Pueblo water rights should be quantified by the Federal Reserved Water Rights Doctrine, commonly known as the *Winters Doctrine*, or some other federal Indian law doctrine such as aboriginal title.

As the "original" inhabitants of North America, Indian tribes have certain aboriginal rights, including aboriginal title to the lands that they have historically occupied. Aboriginal title provides to Indian tribes "the exclusive right to occupy the lands and waters used by them and their ancestors before the United States asserted its sovereignty over these areas."[44] In order to establish aboriginal title, a Tribe must show that it actually, exclusively, and continuously used the property since time immemorial.[45] Aboriginal title is a doctrine long-recognized by the courts.[46]

The water rights of most Tribes are based on federal reserved rights because most Tribes relinquished or lost their historic lands through treaties, congressional action, or executive order. In contrast, the lands of the Pueblos of New Mexico for the most part were not reserved by treaty, act, or executive order. Rather, Pueblos have continuously occupied their lands since long before arrival of Europeans. The United States recognized fee simple title to the Pueblos by statute in 1858.[47] For these reasons, "[t]he recognized fee title of Pueblos is logically inconsistent with the concept of reserved rights."[48]

*Winters* Rights Generally Do Not Apply to Pueblo Lands

Early in the *Aamodt* adjudication, the court determined the *Winters Doctrine* is generally inapplicable to Pueblo Indian water rights. The issue first arose when the *Aamodt* court made an interlocutory order that Pueblo water uses were controlled by the state law of prior appropriation. On appeal, the Tenth Circuit Court of Appeals reversed, holding that the United States had not relinquished jurisdiction and control over the Pueblos; therefore, their water rights were not controlled by state law.[49] The Tenth Circuit considered the nature of the Pueblos' water right, however, and ruled out the *Winters Doctrine* as a basis for claims on long-held Pueblo lands.

The court observed that the Pueblos held their land and water rights in a distinctly different manner than did most Indian Tribes. The Pueblos did not hold title pursuant to a federal reservation created from public land by treaty, act of Congress, or executive order. Instead, the Pueblos' rights were recognized by prior sovereigns, and the United States continued to recognize these rights by the Treaty of Guadalupe Hidalgo. The court noted that the United States confirmed to the Pueblos fee simple title to their lands by the Act of 1858, but the court distinguished this action as a federal relinquishment of title rather than a reservation of rights. The court observed:

> A relinquishment of title by the United States differs from the creation of a reservation for the Indians. In its relinquishment the United States reserved nothing and expressly provided that its action did not affect then existing adverse rights. The mentioned decisions recognizing reserved water rights on reservations created by the United States are not technically applicable.[50]

The court concluded that the "recognized fee title of the Pueblos is logically inconsistent with the concept of a reserved right."[51]

*Aamodt II*: Pueblo Water Rights and Aboriginal Title

With the guidance of the Court of Appeals and its own prior conclusion that *Winters Doctrine* provided no basis for Pueblo water rights, the *Aamodt* adjudication court embarked on an in-depth analysis of the treatment of Pueblo water rights under Spanish and Mexican law and then under the laws of the United States and the Territory and State of New Mexico. The court analyzed the nature of aboriginal water rights and the impact of Spanish, Mexican, and American sovereignty on those water rights.

After extensive analysis of Pueblo water rights under Spanish and Mexican law, the adjudication court concluded that the Pueblos' claims to land and water traditionally used by them were never extinguished by the Spanish or Mexican sovereigns.[52] Rather, the court found these sovereigns "recognized and protected" Pueblo water rights and recognized a right in the Pueblos to increase their water usage based on need.

The adjudication court moved on to consider the effect of United States law on Pueblo water rights, beginning with an analysis of the Treaty of Guadalupe Hidalgo. In its opinion in *Aamodt II*, the court concluded that the Pueblos "came into the United States possessed in fee simple of the property which they occupied, . . . restricted only in the power of alienation and subject to termination by Congress."[53] In reaching this conclusion, the court reasoned that at the time of the Treaty the Pueblos were in full possession of their property and had been for a long period of time, and that this possession was recognized by both Spanish and Mexican governments, subject only to sovereign control of alienation and a sovereign right to sever the property. The Treaty of Guadalupe Hidalgo preserved this status.[54]

The *Aamodt II* opinion reviewed statutory law to determine whether Pueblo title to property and water had been extinguished or diminished under United States law.[55] The court concluded that only the Pueblo Lands Act of 1924 and the Compensation Act of 1933 limited the Pueblos' aboriginal water right. The Pueblo Lands Act of 1924 had the effect of fixing Pueblo water rights to acreage irrigated as of the date of the Act.

The court then reviewed the case law supporting recognition of aboriginal rights and concluded that the Pueblos had aboriginal title to their lands and waters.

> These rights have continued up and until the present day, saving and excepting those terminated by the 1924 Pueblo Lands Act, *supra*. While the 1924 Act provided for termination of the Pueblos' ownership of specific lands within the Pueblos' and termination of the right to use of the water of the stream system thereon, it did not terminate the prior rights of the Pueblos to the use of the water on their remaining lands.[56]

The *Aamodt II* opinion determined the Pueblos' water rights to be as follows:

> The Pueblos have the prior right to use all of the water of the stream

system necessary for their domestic uses and that necessary to irrigate their lands, saving and excepting the land ownership and appurtenant water rights terminated by the operation of the 1924 Pueblo Lands Act, supra. The acreage to which this priority applies is all acreage irrigated by the Pueblos between 1846 and 1924. Acreage under irrigation in 1846 was protected by federal law including the Treaty of Guadalupe Hidalgo, *supra*, and the 1851 Trade and Intercourse Act, *supra*. The Pueblo aboriginal water right, as modified by Spanish and Mexican law, included the right to irrigate new land in response to need. Acreage brought under irrigation between 1846 and 1924 was thus also protected by federal law. The 1924 Act, which gave non-Pueblos within the Pueblo four-square-leagues [about 17,700 acres] their first legal water rights, also fixed the measure of Pueblo water rights to acreage irrigated as of that date.

The nonPueblos' priorities begin as of the date they applied water to the land they used or occupied and which have not been lost by nonuse pursuant to the law of Spain or Mexico or the Territory or State of New Mexico.

The Pueblo water rights appurtenant to their lands are the surface waters of the stream systems and the ground water physically interrelated to the surface water as an integral part of the hydrologic cycle. *Cappaert v. U.S.*, 426 U.S. 128 142 (1976). The Pueblos have the prior right to the use of this water.[57]

In essence, the court determined that the Pueblos have the first priority water right in the stream system for domestic uses and to irrigate all acreage under irrigation as of 1846 and new acreage irrigated between 1846 and 1924. The date of 1846 is significant because it marks the time of United States' occupation of the New Mexico Territory. The court found that the Pueblos' water right as of 1846 was fully protected by federal law and the Treaty of Guadalupe Hidalgo. The period between 1846 and 1924 is significant because the Pueblos had an expanding right based on the Spanish and Mexican law, as preserved by the Treaty of Guadalupe Hidalgo. However, in the court's view, this expanding right terminated and became fixed with the enactment of the Pueblo Lands Act of 1924.[58]

### Clarification and Amplification of *Aamodt II*

The adjudication court's ruling in *Aamodt II* provided a starting point for determining the Pueblos' water rights; however, over the course of further proceedings in the case, questions arose concerning priority and quantification of Pueblo rights. Both questions of priority and quantity relate to whether the water rights claimed are based on aboriginal title. Although the court appeared to have resolved the question by finding in *Aamodt II* that aboriginal title provided the basis for Pueblo water rights on grant lands, the court later recognized that the Pueblos may have reserved rights on reserved lands and might otherwise acquire rights under state law.[59] The court also recognized that certain lands might have a claim under more than one of the three bases for establishing a claim. Depending on the classification of the lands in question as aboriginal, reserved, or other lands, the priority could be different as well as the method used for quantifying the right.[60]

Quantification of Historically Irrigated Acreage. The quantity of an aboriginal irrigation right is based on the "historically irrigated acreage" (HIA).[61] HIA is generally a smaller quantity but has a more senior priority than what could be granted under the practicably irrigable acreage (PIA) standard for reserved rights. Aboriginal rights must be based on actual occupancy of the land since time immemorial.[62] Occupancy, however, is only prima facie evidence of an aboriginal water right. Quantification of an aboriginal irrigation right must be based on actual irrigation between 1846 and 1924.[63]

Domestic and Livestock Uses. The court construed the term "domestic" broadly to include commercial and industrial uses and generally applied the same rules as for irrigation rights. A domestic right is quantified by the maximum water actually used by the Pueblo between 1846 and 1924.[64]

Pueblos may also have water rights for livestock. There is no aboriginal water right for livestock because the Pueblos did not raise livestock until the Spanish introduced it. Therefore, the priority date for livestock use is the date of first use. The quantity of the right is based on historic, customary, and actual use.[65]

### Non-Pueblo Water Rights

The adjudication court's rulings limited the quantity of Pueblo claims, thereby leaving water for other water users in the basin but only after satisfying the Pueblo's senior right. In holding that aboriginal title gave the Pueblos first priority, the court rejected arguments of non-Pueblo ditches or acequias, most of which were established under Spanish and Mexican sovereignty, that water should be allocated under a *repartimiento* or equitable sharing system.[66] In a further nod to the prior appropriation doctrine, the court entered an order limiting all domestic wells permitted after January 13, 1983 to "in house" use only.[67] Such junior wells, known as "post moratorium" wells can only be used for outside uses if a surface water right is transferred to the well.

**Aamodt Parties Reach Settlement**

The adjudication court's holdings made no party a winner. On the one hand the court recognized the first right in the Pueblos but on the other limited the quantity to maximum historical use between 1846 and 1924. Rather than continuing with litigation and numerous ensuing appeals, many of the parties decided to pursue settlement of the Pueblo claims. In 2000, the court ordered the parties to commence settlement negotiations under the supervision of a court-appointed mediator, retired Arizona state court Judge Michael C. Nelson.

Negotiation of Settlement Agreement

Three fundamental settlement principles guided the negotiations. First, the Pueblos would agree to limit in-basin water rights to historical amounts as quantified by the court. Second, the Pueblos would limit priority calls against junior uses in order to protect the status quo within the basin. Third, the federal government would construct a regional water system supplied from the nearby Rio Grande to provide water for additional and future Pueblo uses and to provide an alternative supply to non-Pueblo groundwater pumpers with the aim of reducing demands on the local aquifer.

Over the course of settlement discussions one of the most difficult issues was resolution of Pueblo claims against the approximately 3,000 existing domestic groundwater wells in the basin. Under strict priority administration such wells would be most vulnerable to curtailment in order to protect senior surface flows from depletions caused by groundwater pumping. Nonetheless, because settlements are often not merely based on legal principles and involve other factors such as political and economic considerations, a settlement that did not accommodate a means of domestic supply would not succeed.

In 2004, the negotiating parties announced a draft settlement agreement that called for supply of domestic uses from the regional water system and would have required domestic well owners within the system's service area to discontinue use of their wells. After many well owners voiced opposition to the proposal, the negotiators went back to work, joined by representatives of concerned well owners. Additional settlement discussions required another two years and culminated in a new proposed Settlement Agreement that made connection to the water system voluntary for existing domestic well owners.

On May 3, 2006, a number of parties, including the four Pueblos, the State of New Mexico, the County and City of Santa Fe signed the Settlement Agreement with the support of the two largest associations representing acequias in the basin.[68] Even though the revised agreement made connection to the water system optional, a number of domestic well owners continued to oppose the settlement, arguing the water system would be too costly and should be "Pueblo only," meaning no service should be offered to non-Pueblo residents of the Pojoaque valley.[69] The major settling parties rejected that position. Because availability of water from the system for both Pueblo and non-Pueblo residents remained a key element of settlement, the revised agreement continued to call for a regional system that would afford service to any resident within the system service area.

A complicated document reflecting the many difficult water allocation issues in the basin, the revised Settlement Agreement is based on the following key provisions:

> The Pueblos will forbear from making priority calls in excess of the Pueblos' existing water uses against non-Pueblo surface water users. That means of a total first priority right of 3,660 acre-feet per year (afy), the Pueblos will assert priority for only 1,391 afy.

> Domestic well owners will have a choice of connecting to the water system or keeping their wells. The Pueblos will not make any priority calls against domestic well users provided they agree to one of two options: (a) to connect to the regional water system either once the system is available or subsequently after a change in land ownership; or (b) to keep their wells and not connect to the system, provided each household with annual use in excess of 0.3 to 0.5 afy (depending the category of well) agrees to reduce its use by ten to fifteen percent, or be subject to a priority against the excess amount.

The regional water system will be constructed to deliver treated water from the Rio Grande to the four Pueblos and, to the extent there is sufficient demand in various locations within the basin, to those non-Pueblo water users who elect to receive domestic supply from the water system. Santa Fe County will operate that portion of the water system providing service off of Pueblo lands.

Water users will not be required to cap their wells unless they choose to connect to the water system.

The United States will acquire annual water rights for 2,500 afy of supply diverted from the Rio Grande into the Basin for use by the four Pueblos from the water system, intended, in part, to compensate the Pueblos for their agreement to not fully exercise their right to call priority within the Basin.

Santa Fe County will acquire annual water rights for up to 1,500 afy of supply diverted from the Rio Grande for use by non-Pueblo water users in the Basin.[70]

In addition to finally recognizing and adjudicating Pueblo water rights, the Office of the State Engineer succinctly described the purpose of the Settlement as resolving "conflicts between Pueblo and non-Pueblo water users," protecting existing surface users "from having priority enforcement of the full extent of the Pueblos' first priority water rights" and protecting domestic well users "from priority enforcement for uses below specified amounts."[71] Furthermore, through "the mechanisms outlined in the Settlement Agreement, the parties seek to lessen impacts to the aquifer over time while providing greater reliability of supply in a chronically water-short basin."[72]

Approval of Settlement by Congress

Although all the other governmental entities signed the Settlement Agreement in 2006, federal officials from the Department of the Interior and the Department of Justice would not sign until authorized by Congress. New Mexico Senators Pete Domenici and Jeff Bingaman and Congressman Tom Udall introduced settlement legislation in the 110th Congress to approve the settlement agreement.⁷³ Both the Senate Indian Affairs Committee and the House Resources Committee's Subcommittee on Water and Power held hearings on the bill in September of 2008 but the 110th Congress ended later that year without final action on the legislation.

Led by Senator Bingaman, the New Mexico congressional delegation reintroduced the legislation in the 111th Congress.⁷⁴ Senator Bingaman and newly elected Senator Tom Udall sponsored the bill in the Senate and in the House freshmen Congressmen Ben Ray Luján and Martin Heinrich sponsored an identical bill. In addition to approving the settlement, the bill also proposed to authorize federal appropriations to pay for the federal share of the settlement, most notably $106.4 million for construction costs of the proposed regional water system,⁷⁵ estimated in 2007 to total $159.3 million.⁷⁶

On September 9, 2009, the House Subcommittee on Water and Power held another hearing on the legislation. Representatives of the settling parties testified in support of the legislation. Charlie Dorame, Tesuque Pueblo former governor and chairman of an association of the four Pueblos, described how scarcity of surface water in the basin affected the Pueblos' traditional and ceremonial water uses: "Water is essential to our people for basic needs and our survival, but also for its sacred role in Pueblo culture . . . I have seen the Rio Tesuque go dry many times either before it reaches our village or immediately after it passes through our village."⁷⁷ He testified that the centerpiece of the settlement is the proposed regional water system that will bring as much as 4,000 afy of additional water into the basin. He explained the water system not only will bring needed water to the Pueblos but also will provide water to non-Pueblo water users and thereby "reduce stress on the groundwater resources of the Basin."⁷⁸ In conclusion, Chairman Dorame stated:

> [T]he United States' historic failure to protect the Pueblos' lands and water rights adequately for more than 150 years has led directly to today's conflict over scarce water resources. Once enacted, H.R. 3342 will conserve the shared resource responsibly, bring tangible water to Pueblo and non-Pueblo citizens alike, and will ensure a level of certainty for decades in the Pojoaque Basin. Most important to the Four Pueblos, enactment of this settlement legislation will fulfill the United States trust responsibility and ensure that our children, and their children, can continue our traditions for generations to come.⁷⁹

Also testifying in favor of the settlement, New Mexico State Engineer John D'Antonio recounted the many years of protracted litigation and lost opportunities suffered by the residents of the basin.[80] By contrast, the settlement would provide an amicable and fair resolution by recognizing "large first-priority water rights in the Pueblos commensurate with the acreage historically irrigated by them" but at the same time providing "locally-suited mechanisms whereby centuries-old non-Indian uses will be allowed to continue as well as the Pueblo uses."[81] Water for additional Pueblo uses will be imported from the regional water system with "infrastructure locally appropriate to this settlement, with substantial state and local cost share . . . provided to meet specific Pueblo health, safety and economic development needs."[82] His chief counsel, DL Sanders testified that without settlement many junior users would be subject to curtailment of uses in times of shortages: "Irrigation demand of the Pueblos would be entitled to make a priority call for water sufficient to meet their needs. And I think in two of the three tributaries that we are talking about, that would be sufficient to curtail all other uses. And that would include some domestic use."[83]

Testimony by David Ortiz, president of the largest acequia association in the Pojoaque valley reviewed the long tortured history of land and water disputes in the valley and concluded that settlement is preferable:

> But this settlement legislation offers the promise of protecting the acequia culture unique to northern New Mexico. And, just as the settlement will provide the Pueblos with a permanent water supply, it will also preserve the non-Pueblos' existing use of water for irrigation. As such, the settlement, like all good settlements, does no harm, and it goes a long way to ensure the future of the acequia culture.[84]

He further testified that the regional water system proposed by the settlement would protect household water use in the basin and "over time should reduce ground water withdrawals in the Rio Pojoaque Basin, thereby alleviating stress on the aquifer, and hopefully, restoring surface flows of the rivers in the Basin to levels at which the acequia culture, both Pueblo and non-Pueblo can be permanently sustained."[85]

On behalf of the U.S. Department of the Interior, Commissioner of the Bureau of Reclamation Mike Connor testified in favor of the settlement but stated that full support for the settlement could not be given until the parties addressed a number of concerns, chiefly related to certainty of costs of construction of the regional water system and commitment of non-federal parties to share in those costs.[86] Commissioner Connor commented: "Settlement of the underlying litigation and related claims in this case would fulfill a long-standing federal goal of restoring to the Pueblos the water rights and water resources necessary for their economic and cultural future."[87] And, he continued: "This settlement would accomplish this goal by stabilizing chronic groundwater deficits in the basin without causing harm to local water users."[88] Nonetheless, he requested additional time for Reclamation to work with the settling parties to address the Department of the Interior's remaining concerns.

The only testimony against the settlement came from a group of domestic well owners who continued to object to the regional water system and to harbor concerns that the settlement overall lacked support in the local community.[89]

Santa Fe County Commissioner Harry B. Montoya responded by explaining the benefits of the settlement and the need to have the regional water system available to all county residents, not just Pueblo members:

> Water service should be made available on a non-discriminatory basis to any County resident within the system's service area. I am confounded by the position of some non-Pueblo parties that would deprive other residents of the right to willingly connect. Under the settlement, residents who do not want to connect to the system may keep their domestic wells. Why shouldn't the wishes of other residents who do want to connect also be respected and accommodated?[90]

Recognizing the need to further explain the provisions of the settlement, Commissioner Montoya proposed to conduct a series of "community outreach and settlement focus meetings" in the following months. "The purpose of the meetings will be to hear public concerns and to provide information about the settlement."[91]

In the spring of 2010 Santa Fe County sponsored ten outreach meetings with the assistance of the Stell Water Ombudsman Program, University of New Mexico School of Law. "The meetings were intended to have small audiences, about twenty (20) community members each night, with the goal of encouraging participation."[92] The meetings were successful in explaining the terms of the settlement agreement and fostering a forum for discussion.

During the same time frame, the settling parties worked with the Bureau of Reclamation to address concerns raised by Commissioner Connor in his testimony before Congress. Following the House committee hearing, the Commissioner directed Reclamation staff to undertake a design, estimating and construction (DEC) review of the project costs. Senior Reclamation engineers completed the DEC review at the end of 2009.[93] Their report found no fatal flaws in the proposed plans for the regional water system but emphasized that current cost estimates could not be relied upon until more thorough engineering work is completed. In response the non-federal settling parties agreed to a revision to the legislation making clear that the United States would not be the only party responsible for any cost overruns. The settling parties also agreed to a number of technical changes requested by the Department of the Interior. On September 17, 2010, Commissioner Connor, joined by Department of the Interior counselor Alletta Belin, wrote to the respective congressional committees expressing the Department's endorsement of the legislation, with the agreed revisions.[94]

As the 111th Congress began its second session in 2010, prospects for the *Aamodt* settlement legislation looked promising. Shortly after the House Committee on Natural Resources favorably reported the bill, the House passed it by a vote of 249 to 153 on January 21, 2010, and sent it to the Senate for action. But there it languished, as run-up to mid-term elections and Senate voting rules afforded very little floor time. With the 111th Congress winding down after contentious mid-term elections, the settling parties feared yet another Congress would end without passage of the legislation. At the direction of Senator Bingaman, Senate Energy and Natural Resources Committee lawyer Tanya Trujillo worked skillfully to position the bill for passage during the lame-duck session. On November 19, 2010, as Senator Bingaman presided over the Senate the bill unanimously passed the Senate as part of the Claims Resolution Act of 2010, which included the Taos Pueblo water rights settlement, two other Indian water rights settlements and settlement of two big class action cases, *Cobell v. Salazar* and *Pigford v. Glickman.* Promptly the House passed the combined bill 256 to 152 on November 30 and the President signed it into law on December 8, 2010, when it became Public Law No. 111-291.[95]

Greatly accentuating this legislative feat, Senator Bingaman succeeded in including substantial appropriations in the bill for both *Aamodt* and Taos settlements and for the previously enacted Navajo settlement. The bill appropriated $81.8 million for the *Aamodt* settlement, including $56.4 million for the regional water system, with an authorization of $92.5 million for the remaining federal share of settlement costs. "With the strong backing of the

Obama administration, we are able to finally bring these long-standing water claims to a positive conclusion. Under these settlements, thousands of New Mexicans will have the certainty about their water rights—a goal that is 40 years in the making," Senator Bingaman stated upon passage by the Senate.[96]

## Conclusion

One hundred years after statehood, the settlement of water disputes in the Pojoaque Valley should give residents reason for optimism. Like all general stream adjudications and other major water cases, allocation and usage disputes never really end. As long as people continue to live together in shared water basins, there will be disagreements about the rules and how they apply. However, once the adjudication court approves the *Aamodt* settlement and after the water infrastructure and other settlement features are completed, the argument will dramatically change. No longer will the debate focus on who claims what. Instead, the challenge to residents of the Pojoaque Valley will be to live within their respective rights and to work together to support the institutions and mechanisms established to protect their shared resource.

## Notes

1. The nineteen Pueblos of New Mexico are the Pueblos of Taos, Picuris, Ohkay Owingeh (formerly San Juan Pueblo), Santa Clara, San Ildefonso, Tesuque, Nambe, Pojoaque, Cochiti, Santo Domingo, San Filipe, Santa Ana, Zia, Jemez, Sandia, Isleta, Laguna, Acoma, and Zuni. All but the Zuni Pueblo are located along the Rio Grande and its tributaries. Dan Scurlock, U.S. Department of Agriculture, General Technical Report RMRS-GTR-5, *From the Rio to the Sierra: An Environmental History of the Middle Rio Grande Basin* 84–85 (1998).
2. *Id.* at 84 (1998). The Pueblos of San Ildefonso, Tesuque, Nambe, and Pojoaque exclusively occupied their lands since prior to the 14th Century A.D. *Aamodt II* at 996.
3. *Zuni Tribe of N.M. v. United States*, 12 Ct. Cl. 607, 613 (1987). This case provides a detailed history of the Zuni Pueblo.
4. Clifford S. Crawford, et al., *Middle Rio Grande Ecosystem: Bosque Biological Management Plan* 23 (1993) (on file with the U.S. Fish & Wildlife Service in Albuquerque, NM).
5. *New Mexico v. Aamodt*, 537 F.2d 1102, 1105 (10th cir. 1976) [hereinafter *Aamodt I*]; *New Mexico v. Aamodt*, 618 F. Supp. 993, 996 [hereinafter *Aamodt II*].
6. There is uncertainty whether the Pueblos held their Pueblos pursuant to Spanish grants or whether the Spanish government merely recognized and protected Pueblo title. *Compare Mountain States Telephone & Telegraph Co. v. Pueblo of Santa Ana*, 472 U.S. 237, 240 (1985) (Pueblo land ownership recognized by grants from the King of Spain); *United States v. Sandoval*, 231 U.S. 28, 38 (1913) (Pueblos held lands in communal, fee-simple ownership

under grants from the King of Spain) & *United States v. Joseph*, 94 U.S. 614, 618 (1876) (Pueblo title dates back to grants made by the government of Spain) *with Aamodt II*, 618 F. Supp. at 997 (claims to royal grants are based on oral tradition and no physical evidence exists of grants from the Spanish crown).
7. *Aamodt II* at 998.
8. *Id.* at 998.
9. *Id.* at 999.
10. *Id.*
11. *Id.*
12. One difference was that Spain treated the Pueblos as wards, whereas Mexico treated them as citizens. *Id.* at 1000.
13. *Id.* at 996.
14. *Id.* at 998. In this regard the Special Master's Conclusions of Law as adopted by the court in *Aamodt II* appear to conflict. Conclusion Number 4 states that both Spain and Mexico recognized and protected a "prior and paramount right" in the Pueblos, while Conclusion Number 16 states that Mexican law "did not give any preference, or prior or paramount interest, to Indian irrigation needs." *Id.*
15. Treaty of Peace between the United States and Mexico, Feb. 2, 1848, Preamble, NMSA 1978, chap. 1, Pamp. 3; 9 Stat. 922. [hereinafter Treaty or Treaty of Guadalupe Hidalgo].
16. *Id.* art. V. The international boundary was later modified by the "Gadsden Purchase." Gadsden Treaty between the United States and Mexico, Dec. 30, 1853, NMSA 1978, chap. 1, Pamp. 3; 10 Stat. 1031)
17. Treaty of Guadalupe Hidalgo art. VIII.
18. *Id.*
19. *Aamodt II* at 1000; *Zuni Tribe of Indians v. United States*, 12 Cl. Ct. 607, 627 (1987); *United States v. Lucero*, 1 N.M. 4301-35 (Jan. Term 1869).
20. *Aamodt II* at 1001.
21. 4 Stat. 730, § 12, 25 U.S.C. § 177.
22. Trade and Nonintercourse Act of 1851, 9 Stat. 574, 587.
23. 11 Stat. 374.
24. *Id.*
25. *Mountain States*, 472 U.S. at 243 & n.14 (citing S. Rep. No. 492, 68[th] Cong., 1[st] Sess., 5 (1924).
26. *United States v. Joseph*, 94 U.S. 614 (1876).
27. Some non-Indian acquisitions of Pueblo lands were by good-faith conveyance from the Pueblos. Other acquisitions may not have been in good faith. *Aamodt II* at 1002. Congressional hearings in 1924 estimated that 80 percent of the non-Indian claims to Pueblo lands were not resisted by the Pueblos. *Mountain States*, 472 U.S. at 244 n.14.
28. *Id.* at 618-19.
29. *Id.*
30. *United States v. Lucero*, 1 N.M. 422 (1869). The *Joseph* decision affirmed a similar case that was originally brought in the territorial court in *United States v. Varela*, 1 N.M. 593 (N.M. Terr. Jan. Term 1874).

31. *Pueblo of Nambe v. Romero*, 10 N.M. 58 (1900).
32. 14 N.M. 1 (1907). The Act of January 30, 1897, 29 Stat. 506, ch. 109 made it unlawful to introduce intoxicating liquor into Indian Country.
33. New Mexico Enabling Act, June 20, 1910, 36 Stat. 557.
34. *United States v. Sandoval*, 231 U.S. 28, (1913).
35. *Sandoval*, 231 U.S. at 39, 47.
36. *Id.* at 48-49.
37. *United States v. Candelaria*, 271 U.S. 432, 441-42 (1926).
38. *Mountain States*, 472 U.S. 237 (1985) (citing S. Rep. No. 492, 68th Cong., 1st Sess., 5) (1924).
39. Pueblo Lands Act of 1924; *Aamodt II* at 1004.
40. Pueblo Lands Act of 1924, § 3.
41. Pueblo water rights are being resolved in a number of ongoing stream system adjudications in New Mexico. Seven other ongoing water rights adjudications in New Mexico involve Pueblo Indian water rights: *New Mexico v. Abbott*, Nos. CIV-7488C & CIV-8650C (D.N.M.) (Santa Clara and San Ildefonso Pueblos and Ohkay Owingeh (formerly San Juan Pueblo)); *New Mexico v. Abeyta*, Nos. CIV-7896C & CIV-7939C (D.N.M.) (Taos Pueblo); *New Mexico v. Aragon*, No. CIV-7941C (D.N.M.) (Ohkay Owingeh); *United States v. Abousleman*, No. CIV-83-1041C (D.N.M.) (Jemez, Santa Ana and Zia Pueblos); *Anaya v. Public Service Company*, Santa Fe Co., No. 43,347 (Cochiti Pueblo); *New Mexico v. Kerr-McGee Corp.*, Cibola Co., Nos. CB-83-190CV & CB-83-220-CV (Acoma and Laguna Pueblos; *Gallup v. United States*, McKinley Co., No. CV-84-164 (Zuni Pueblo). Ed Newville, Comment, *Pueblo Indian Water Rights: Overview and Update on the Aamodt Litigation*, 29 Nat. Resources J. 251, 277 n.91 (Winter 1989).
42. No. 00639 (D.N.M.). The Court of Claims extensively analyzed aboriginal title of the Zuni Pueblo in the context of land claims rather than water claims. The analysis was similar to the historical analysis in *Aamodt II*. *Zuni Tribe of N.M. v. United States*, 12 Cl. Ct. 607 (1987); *Zuni Tribe of N.M. v. United States*, 12 Cl. Ct. 641 (1987).
43. *Aamodt II* at 995.
44. *Yankton Sioux Tribe of Indians v. South Dakota*, 796 F.2d 241, 243 (8th Cir. 1986) (citing People of the Village of Gambell v. Clark, 746 F.2d 572, 574 (9th Cir.1984); *Oneida Indian Nation v. County of Oneida*, 414 U.S. 661, 66769 (1974).
45. *United States v. Santa Fe Pacific Railroad Co.*, 314 U.S.339, 345 (1941); *Sac & Fox Tribe of Indians v. United States*, 179 Ct. Cl. 8, 383 F.2d 991, 998 (1967); *Yankton Sioux Tribe of Indians v. South Dakota*, 796 F.2d 291 (8th Cir. 1986); Adair, 723 F.2d at 1414; *State ex rel. Martinez v. Lewis*, 861 P.2d 235, 255, 116 N.M. 194, 214 (N.M. App. 1993).
46. For a discussion of the historical development of the theory of aboriginal title, see Charles T. DuMars, Marilyn O'Leary & Albert E. Utton, *Pueblo Indian Water Rights* 11-16 (1984).
47. Act of 1858, 11 stat. 374.
48. *Aamodt I* at 1111. The court explained: "[A] relinquishment of title by the United States differs from the creation of a reservation for the Indians. In its relinquishment, the United States reserved nothing and expressly provided that its action did not affect then existing adverse rights. The mentioned decisions recognizing reserved water rights on reservations created by the United States are not technically applicable." *Id.*
49. *Id.* at 1104, 1111.

50. *Id.* at 1111; *State ex rel. Reynolds v. Aamodt*, No. CIV-6639 (Mem. Op. & Order, June 10, 1983, at 4–6 (Doc. No. 728)).
51. *Id.* The adjudication court agreed for the same reasons that the *Winters Doctrine* did not provide a basis for Pueblo water rights. *Aamodt*, No. CIV-6639 (Mem Op & Order, June 10, 1983, at 9–10 (Doc. No. 728)). The federal district court agreed although it did not believe that the Court of Appeals' determination of Pueblo rights was necessary to its determination of the issue before it, and therefore not the "law of the case." The adjudication court thus concluded that the Pueblo lands, with the exception of certain lands set aside by Executive Order for the Nambe Pueblo, did not have *Winters* rights. *Aamodt II* at 1010.
52. *Id.* at 998.
53. *Id.* at 1001.
54. *Id.* at 1000.
55. The court reviewed the following authorities: United States Const. art 1, § 8 (Congress has the power to regulate commerce with the Indian tribes); Trade and Nonintercourse Act of 1834, 4 Stat. 730, § 12, 25 U.S.C. § 177 (alienation of Indian lands is only valid if made by treaty or constitutional convention); Confirmation Act of 1858, 11 Stat. 374 (confirming the land claims of certain Pueblos); Enabling Act for New Mexico, June 20, 1910, 36 Stat. 557, § 2 (the term Indian Country includes the Pueblo Indians of New Mexico and such lands are "subject to the disposition and under the absolute control of the Congress"); Pueblo Lands Act of 1924, ch. 331, § 1 *et seq.*, 43 Stat. 636 (creating Pueblo Lands Board to quiet title to Pueblo lands); Pueblo Compensation Act of 1933, 48 Stat. 108 (compensating Pueblos and non-Pueblo settlers who lost land under the 1924 Act).
56. *Id.* at 1009. Whether an aboriginal right is better than a *Winters* right depends on the circumstances. An aboriginal right is better where priority is the concern because an aboriginal right is the senior right on the stream system. A *Winters* right, on the other hand, has a priority date as of the date of the reservation. If quantity of water is the concern, then a *Winters* right may be preferable because the quantity of the right is based on the "practicably irrigable acreage" (PIA) standard, which may be greater than the acreage irrigated between 1846 and 1924.
57. *Id.* at 1010. In a later memorandum opinion and order, the adjudication court recited its ruling with respect to the 1924 and 1933 acts but provided little additional insight into its reasoning. The court merely stated that with passage of the 1924 and 1933 acts, Congress took "affirmative political action" to end the controversy regarding the Pueblos' land and water rights by quieting title to all rights developed prior to 1924, and that this had the effect to fix the Pueblo rights. *Aamodt*, No. CIV-6639 (Mem. Op. & Order, Jan. 31, 2001, at 4 (Doc. No. 5642)) (citing *Mountain States*, 472 U.S. at 240, 244; *Aamodt I*, at 1105-06; *Aamodt II*, at 1004). This may have been a practical recognition that water rights within the basin were fully appropriated and an expanding Indian water rights claim would continually conflict with non-Indian water rights claims.
58. Consequently, the court determined that the Pueblos have a prior right to use all of the water of the stream system necessary for domestic and irrigation needs on all acreage irrigated by the Pueblos between 1846 and 1924, except for lands lost pursuant to the 1924 Act. The court reasoned that irrigated acreage as of 1846 was an aboriginal right protected

by the Treaty of Guadalupe Hidalgo (preserving the rights of Mexicans within the ceded territory) and the 1851 Trade and Intercourse Act (prohibiting alienation of Indian lands without congressional consent). The court further reasoned that the Pueblo aboriginal water right, "as modified by Spanish and Mexican law," included the right to irrigate new acreage based on need; therefore, the Pueblos had a federally protected right to water for newly irrigated acreage between 1846 and 1924.

59. For instance, the court recognized that the Nambe Pueblo had a *Winters* right claim for lands set aside by a 1902 Executive Order. *Aamodt II* at 1010.

60. The court's ruling in *Aamodt II* that the Pueblos had aboriginal claims to water rights was perhaps an oversimplification of the nature of Pueblo claims. One of the first problems is dealing with Pueblo water rights claims for lands that were reacquired or replaced through operation of the Pueblo Lands Act. This problem led the court to develop a classification scheme that includes "grant lands," "acquired lands," "reacquired lands" and "replacement lands."

The court used the term "grant lands" to refer to lands granted by prior sovereign that the Pueblos have continuously occupied since time immemorial. This is the simplest case. The Pueblos have aboriginal water rights claims for water appurtenant to these grant lands. *Aamodt*, No. CIV-6639 (Mem. Op & Order, Apr. 14, 2000 at 8-9 (Doc. No. 5596)). Water rights established for grant lands have a "time immemorial" priority.

"Reacquired lands" are lands within the exterior boundary of a Pueblo grant that the Pueblo lost to non-Pueblo claimants as a result of the Pueblo Act of 1924 but then later reacquired by purchase. These lands also support aboriginal water rights claims. The court reasoned that the 1924 Act extinguished Pueblo land claims but not water rights claims. When land passed to a non-Pueblo owner by operation of the Act, a Pueblo's aboriginal title to water rights merely became dormant. Aboriginal title is reactivated when the Pueblo reacquires the land. *Aamodt*, No. CIV-6639 (Mem. Op & Order, Apr. 14, 2000, at 8 (Doc. No. 5596)). In enacting the Pueblo Lands Act of 1924 and the Compensation Act of 1933, Congress intended for the lands acquired under the acts to replace what the Pueblos had lost, and for these lands be equivalent to the lost lands, particularly with respect to water rights. *Id.* (citing S. Rep. No. 492 on S. 2932, 68th Cong., 1st Sess. At 7-8 (Apr. 24, 1924)). Liberally construing congressional intent, the Pueblos also have time immemorial priority for water rights appurtenant to lands that they lost by operation of the 1924 Act and then reacquired. *Id.* at 5.

"Acquired lands" are lands that a Pueblo acquired outside the exterior boundaries of its grant. A Pueblo may claim aboriginal title to water rights if it can prove that it historically irrigated this acreage. If not, the Pueblo may base a claim under state law. *Id.* at 8-9.

"Replacement lands" are lands acquired by a Pueblo under the 1924 Act that are not within the Pueblo's grant boundary or aboriginal territory, or that are located within the grant boundary or aboriginal territory but for which the Pueblo cannot prove historic irrigation. *Aamodt*, No. CIV-6639 (Mem. Op & Order, Apr. 14, 2000, at 9 (Doc. No. 5596)). It is not clear how the *Aamodt* court intended to treat priority for replacement lands. In a 1987 Memorandum Opinion and Order, the court indicated that replacement lands acquired by operation of the Pueblo Land Act of 1924 would have immemorial water rights. However, at that time the court appeared to equate replacement lands with acquired lands. The definition of replacement lands as set forth in the court's April 2000 Memorandum is inconsistent with the definition of acquired lands. This issue remained unresolved.

61. *Aamodt*, No. CIV-6639 (Mem. Op & Order, Dec. 29, 1993 at 2 (Doc. No. 4264)).
62. *Aamodt*, No. CIV-6639 (Mem. Op & Order, Dec. 29, 1993 at 2 (Doc. No. 4264)) (citing *Yankton Sioux Tribe of Indians v. South Dakota*, 796 F.2d 241, 243 (8th cir. 1986)); *Oneida* 223-34 (1985); *Santa Fe R.*, at 345, 348; *United States v. Pueblo of San Ildefonso*, 206 Ct. Cl. 649, 513 F.2d 1383, 1394 (1975).
63. *Aamodt*, No. CIV-6639 (Mem. Op. & Order, Feb. 26, 1987 at 10 (Doc. No. 2977)) (citing *Aamodt I* at 1010).
64. *Aamodt*, No. CIV-6639 (Mem. Op. & Order, Jan. 31, 2001, at 10 (Doc. No. 5642)).
65. *Id.* at 5, 11.
66. Mark F. Sheridan, *Pueblo Indian Water Rights, the Federal Law Sources, A Non-Pueblo Position* (Jan. 2002) (CLE International, Law of the Rio Grande conference).
67. *Aamodt*, No. CIV-6639 (Order, Jan. 13, 1983).
68. The Rio Pojoaque Acequia and Water Well Association, Inc, represented by Mark F. Sheridan, and the Rio de Tesuque Acequia Association, represented by Larry C. White.
69. Written testimony by Richard Rochester, President Pojoaque Basin Water Alliance and Vicente Roybal, submitted September 9, 2009 to the Committee on Natural Resources, Subcommittee on Water and Power, U.S. House of Representatives, hearing of H.R. 3342, Aamodt Litigation Settlement Act.
70. *Overview of the Revised Aamodt Settlement Agreement*, 2006, prepared by New Mexico Office of the State Engineer.
71. *Id.*
72. *Id.*
73. Aamodt Litigation Settlement Act, S. 3381 and H.R. 6768, 110th Congress, 2D Session.
74. Aamodt Litigation Settlement Act, S. 1105 and H.R. 3342, 111th Congress, 1st Session.
75. *Id.* at Sec. 107(a)(1).
76. The difference of $52.9 million was the state and local cost share estimated in 2007, not including the cost of customer service connections.
77. Statement of The Honorable Charles J. Dorame, Chairman, Northern Pueblos Tributary Water Rights Association, and Former Governor, Pueblo of Tesuque before the House Committee on Natural Resources, Subcommittee on Water and Power, H.R. 3342, Aamodt Litigation Settlement Act, September 9, 2009, Serial No. 111-34, page 34.
78. *Id.* at 35.
79. *Id.* at 36.
80. Statement of John D'Antonio, New Mexico State Engineer, before the House Committee on Natural Resources, Subcommittee on Water and Power, H.R. 3342, Aamodt Litigation Settlement Act, September 9, 2009, Serial No. 111-34, page 42.
81. *Id.*
82. *Id.*
83. Testimony of DL Sanders, Chief Counsel, New Mexico State Engineer, before the House Committee on Natural Resources, Subcommittee on Water and Power, H.R. 3342, Aamodt

Litigation Settlement Act, September 9, 2009, Serial No. 111-34, page 55.
84. Statement of J. David Ortiz, President of Rio Pojoaque Acequia and Water Well Association, Inc., Hearing before the House Committee on Natural Resources, Subcommittee on Water and Power, H.R. 3342, Aamodt Litigation Settlement Act, September 9, 2009, Serial No. 111-34, pp. 4–5.
85. *Id.* at p. 5.
86. Statement of Michael L. Connor, Commissioner Bureau of Reclamation, U.S. Department of the Interior, Hearing before the Committee on Natural Resources, Subcommittee on Water and Power, U.S. House of Representatives on H.R. 3342, September 9, 2009, Serial No. 111-34, p. 18
87. *Id.* at p. 22.
88. *Id.*
89. Testimony in Opposition by Richard Rochester, President Pojoaque Basin Water Alliance and Vicente Roybal, submitted September 9, 2009 to House Committee on Natural Resources, Subcommittee on Water and Power, H.R. 3342, Aamodt Litigation Settlement Act, Serial No. 111-34, p. 75.
90. Supplemental Statement of Harry B. Montoya, Santa Fe County Commissioner, New Mexico to the House Committee on Natural Resources, Subcommittee on Water and Power, H.R. 3342, Aamodt Litigation Settlement Act, September 23, 2009, Serial No. 111-34, p. 70
91. *Id.* at p. 71.
92. Aamodt Settlement Outreach Report, May 2010, Darcy S. Bushnell, Program Director, Joe M. Stell Ombudsman Program, Utton Transboundary Resources Center, University of New Mexico School of Law, p. 1.
93. Design, Estimating, and Construction Review, Pojoaque Basin Regional Water System, Bureau of Reclamation, Upper-Colorado Region, New Mexico, December 2009.
94. Letter from Michael L. Connor, Commissioner, Bureau of Reclamation, and Alletta Belin, Counselor to the Deputy Secretary of the Interior, to The Honorable Byron Dorgan Chairman, Committee on Indian Affairs, The Honorable Grace Napolitano Chairwoman, Subcommittee on Water and Power, and other committee members, dated September 17, 2010.
95. Claims Resolution Act of 2010, Aamodt Litigation Settlement Act, Title VI, Sec, 617, Public Law No. 111-291.
96. Senate approves Aamodt settlement, funding for water system, *Santa Fe New Mexico*, November 19, 2010.

**Select References**

Peter C. Chestnut, *Pueblo Water Rights*, Overview of Water Law Applicable to
The Middle Rio Grande Water Planning Region, January 2003.

Mark F. Sheridan, *Pueblo Indian Water Rights, the Federal Law Sources, A Non-Pueblo Position* (Jan. 2002) (CLE International, Law of the Rio Grande conference).

# *Aamodt, Schmaamodt: Who Really Gets the Water?*

### John Nichols

Note: John Nichols gave a version of the following talk in Nambe, New Mexico on July 18, 2004. The paper was published in the Taos News and the Santa Fe New Mexican, in a collection of essays, *To Harvest, To Hunt: Stories of Resource Use in the American West*, and will appear in a book of essays produced by the Taos Historical Society in 2010. The version here is by permission of the author.

When I arrived in Taos thirty-five years ago I bought a little adobe house in the Upper Ranchitos part of town. It was on an acre and a half of land that was fed by two irrigation ditches. I came from the East, and I knew absolutely nothing about acequias or water laws in New Mexico. The only things I had going for me were that I could speak Spanish after a fashion and I was eager to learn.

I had been in my new house for about two weeks when I got a letter from the New Mexico State Engineer's Office. It was an offer of judgment on the water rights of my land, and was part of a massive adjudication suit that is still going on today. Back then I had no idea what the state engineer was talking about, and of course my heart fell into my toes because I thought I was being sued in some way that would tangle me up in lawyers, land me in jail, and take me away from the land that I had owned for about five minutes, total.

Welcome to Nuevo México, tierra del encanto.

About two minutes after my heart fell into my toes, the mayordomo of the Pacheco Ditch, Eloy Pacheco, showed up at my house to inform me that all the parciantes were going to clean that ditch on Saturday. So, I went into town and bought a shovel, and showed up on Saturday to clean the acequia. I had never worked on a chain gang before, so I had never worked so hard in my life. It wasn't a very long or difficult ditch, but there were about 40 of us who worked hard all day long cleaning up that artery, plugging the perrito holes, burning the grass, chopping out jaras, and cursing the viejitos who just leaned on their shovels shouting "Vuelta!" every few minutes and babbling to each other about how hard they used to work in the good old days, when men were really men and the acequias truly meant something. I was the only gavacho on the crew, and that was the first time I met all my neighbors together, and I had a blast, complaining, bitching about the work, listening to filthy jokes, and to all the chisme and mitote of the neighborhood and Taos. I picked up a lot of history, too. Stories galore. And plenty of laughter.

The old guys are mostly all gone now. Adolfo Lavadie, Alfonso Tejada, Eloy Pacheco. Phil Miera is still kicking, however. The young people like Jerry Pacheco and his brother, Bobby and Joe Cordova are getting older like me. I'll be 64 next week. But the acequias are still running with water every spring and summer . . . and the adjudication will eventually run its course, but it may cause considerable damage to the acequia system that created the valley in the first place.

Water is what connected me to my little plot of land, and to my neighbors, to my community, and ultimately to the larger world of New Mexico. There was a half-acre of pasture in front of my house, irrigated by the Pacheco Ditch. The half-acre pasture in back was irrigated by the Lovatos Ditch. Near the house was a small garden that I watered with a hand-dug well beside the house. The house was also connected to a community water users association for domestic use.

The first thing I did in Taos was plant a garden, and for 20 years I had a beautiful garden. Every year I grew corn and green beans and carrots and beets and squash and broccoli and peas and sunflowers and I even tried to grow watermelons a couple of times, but they never got bigger than baseballs. I had more pears and apples than I knew what to do with in the autumn. And I had chickens and turkeys, also. My neighbor, Tom Trujillo, put his horse in my front pasture, and then in my back pasture, and I spent more time talking to that horse than Tom ever did. In the spring and summer I let the grass grow in the two fields, and then Tom would bale it for his animals. I fixed my fences, and cleaned my little irrigation channels, and carefully irrigated all the living space on that property.

It got so that I really loved watering that land. I loved it because it gave me common ground with all my neighbors. We gathered at meetings to discuss the acequias, and during and after those meetings we discussed everything else. We got to know each other. We became friends. I got to palaver with the mayordomos, and that was fun. I treasured friends like Tom Trujillo and Bernardo Trujillo and Eloy Pacheco and Phil Miera. There were lots of fights and tense times. The Lovatos Ditch runs through Pueblo land, and so we were always palavering with people at the Pueblo about ditch problems. In dry summers, when the Pueblo River was really low, we had a heckuva time getting enough water. You could bet that some idiot on the Molina Ditch would divert so much of the river through their compuerta that the Pacheco Ditch would run dry. Then our hotheads would go over and kick out their diversion dam, and the next thing you know they'd be threatening to attack us with rifles.

In a real wet year, with a bad spring runoff, naturally our compuerta would get blown out in a flood. I say compuerta with a grain of salt, because what we called a compuerta was basically a lot of rocks, old tires, down fences and railroad ties, black plastic and chicken wire, with maybe a dead horse thrown in for good measure, all piled at a slant into the river in order to divert some of the flow into the acequia. I would stand waist deep in ice cold water trying to tie down all that stuff with cables and metal fence posts, and I'd snarl over at Alfonso Tejada or Eloy Pacheco or Adolfo Lavadie or Jerry Pacheco, "Ain't there a better way of doing this?" and they'd just shrug and say, "well we've been doing it like this for 400 years. And so far it seems to work." People didn't have very much money, but they knew how to get the job done.

The acequias and my property connected me to those four hundred years. And to the people, culture, and community that had revolved around the water for all that time. They connected me to a tradition rich in grass and orchards and animals and healthy neighborhoods. They connected me to a language rich in history and personality and soul.

Of course, it wasn't always easy to keep those connections healthy. For me, those two acequias that fed water to my property were sometimes one wonderful disaster after another. Some years the muskrats went absolutely postal, and we had to form emergency brigades to try and repair the ditch banks between Tom Trujillo's property and Sebastian's Bar. Occasionally those water dogs would hit my ditch bank like roto-rooters going berserk, and although I'm not the bloodthirsty type, I actually wound up trying to shoot them from my kitchen window before the ditch bank collapsed and drowned Tom Trujillo's horse.

Because we had such a primitive compuerta on the Pacheco Ditch, it wasn't really feasible to kick out the diversion dam, so water often ran all winter. Sometimes it froze low in the ditch. The water rose, and froze again, and rose, and froze again, and pretty soon the ice forced the water over the top of the banks into my front field and Tom Trujillo's front field, and it froze really solid. Then one day all the kids in the neighborhood came skating. They brought with them dozens of dogs. And the dogs immediately attacked my chickens and killed all of them while I was taking a nap. There is no end to the tragedies that can be caused by water.

Sometimes I would leave open a compuerta in my back field and again, at night, somebody would throw the entire Pueblo River into a veinita six inches deep, and when I woke up in the morning the chickens would be roosting on top of their shed, their eggs would be floating in the Kingdom Hall Church parking lot next door, and my outhouse would be transformed into an overflowing septic tank.

When I say "compuerta in the back field" I should explain that for me a "compuerta" in my back field was just a hole I chopped in the side of the ditch with a shovel. It used to amaze me to watch how my neighbors irrigated compared to myself. Me, because my backfield was at the end of the ditch, I always got water last, and usually at night. It was only a half-acre of grass with a lateral going through it. You'd think I could just cut one little opening and flood the field. I mean, that's how my neighbors did it. I'd watch old Adolfo Lavadie, a little bent over grasshopper of a viejito, at his fields. He'd make one little shovel cut, plop the dirt in the proper spot, then lean on his shovel and watch the water calmly flood five acres without a hitch. Me, I'd chop one path, irrigate forty square feet, then chop another path, irrigate another forty square feet, chop a third path, irrigate another forty feet—what was the matter with me? I could never figure out the Euclidian geometry necessary to laying down a perfect sheet of water across the grass. I finally figured out it was the genes cultivated during those past four hundred years of creating and caretaking the acequias that gave my neighbors such an advantage.

I remember once a cow died on Indian land and fell into the Lovatos Ditch and blocked it. But we weren't allowed on Indian land without permission from the Pueblo. So we went out to the Pueblo and the governor's office said they'd send a crew out to get rid of the cow. But they never sent the crew. So we went back to the Pueblo and they told us not to worry, the crew would be there that afternoon. And that afternoon a bunch of us waited on the road outside the fence near the rotting cow, but the Pueblo crew never showed up. Of course, you could bet if we jumped the fence and started shoveling out the cow on our own, that's when the Pueblo crew probably would've showed up and accused us of trespassing on federal land, and we all would have wound up in the government prison at Leavenworth, Kansas. So, we tried once again to speak with the Governor, the War Chief, and who knows who else, and they promised to deal with the cow.

But nothing happened. So finally I went down there with a buddy, we tied kerchiefs over our noses, and shoveled out that cow. I think it was probably the most unpleasant task I ever performed on a New Mexico acequia. And afterwards I figured the Pueblo was probably laughing at us the whole time. I mean, right from the git-go they weren't idiotic enough to place themselves knee deep in bovine gore, and they were probably trying to figure out: What had taken us so long to catch on to that fact and jump the fence and break the law, and get rid of their rotten cow for them?

Of course, one of the most detestable jobs on the acequia is collecting money. The ditch fees. The yearly dues. The work fees to hire peones if the parciante no quiere limpiar la acequia herself . . . or himself. In my experience on the Pacheco Ditch it was always Eloy Pacheco, the mayordomo, who performed this invaluable service. He was pretty informal. He'd known everybody for 70 years and he had his ways. Sometimes he'd get the money, sometimes he didn't. It was a frustrating job. Then I was elected a commissioner on the ditch, and the other commissioners decided to make me treasurer. Thanks a lot, you guys. You could say that the record keeping up until then had been fairly lax, and my orders were to clean it up because the state was really starting to poke its nose into acequia business because of the ongoing adjudication suit. They also decided that I should collect the money. Me? I gulped, said okay, and went about my task with the zeal of a born-again tax collector. I hated the job, I was terrified of the job, but I realized the money was really important to the well-being of the acequia, and also to keeping people involved as modern times began to weaken the commitment many were beginning to feel toward the land.

"Hey, Effie Sebastian," I would say on the phone, "where's our twenty dollars?"

"*A la chingada con your viente pesos,*" Effie would shout back. She's 85 years old. "I don't use the water anymore, so a mi no me importa if I lose it."

"Don't be a pendeja, Effie. Sin derechos de agua tu terreno no vale nada."

"Tú no vales nada, gringo," Effie would shout, and hang up the phone.

So I dialed her again.

"We gotta have that money, Effie. We have to hire a backhoe to dig out a cave-in over by Archie Anglada's."

"Why do I have to pay to dig out a cave-in at Archie Anglada's?"

"Because you are a parciante on the acequia, viejita."

"Who you calling viejita, nene?"

"Please. Por favor. Dios te lo pagará después."

"Tell God he can pay my ditch fees for me now, bobo."

And she'd hang up the phone once more.

Eventually, I got the money. It was just a ritual we were going through. I imagine people have been going through that ritual since time immemorial. Still, as the years passed on it was tougher to get the money. People complained more, they had more excuses, they weren't as enthusiastic, they were starting to lose interest. The old guys on the acequia died off. The families sold their animals. There weren't so many gardens anymore. Newcomers moved in and built houses where there'd once been fields and the newcomers weren't so interested in the water. Each year, when we went to clean the acequia, there were less peones to do the work. Homeowners would not want to do the work, and they would hire young boys, cheap, to do the work, but the boys couldn't really do the work. And it became more difficult to find people willing to work for the prices offered. And some years, while I still had my property on the acequia, when we went to clean the ditch in the springtime we'd only have maybe 15 people. And you can't do a very good job that way. So we'd have to raise fees, collect the money, and hire a backhoe to do part of the job. A backhoe is a wondrous machine that can replace 15, 20, 25 people working together in the same amount of time. But the backhoe is run by a single person, who's usually not even connected to the acequia, to its history, or to the community that for centuries has drawn its lifeblood off that artery. So the backhoe can be the beginning of the end for much of what makes the acequia, and its water, truly valuable. When the backhoe becomes a necessity on a ditch, then it's a good bet the acequia is going to lose its water. Parciantes will sell it to a business, to a hotel, to be used in flushing toilets. And after a while the community on the acequia won't be so interactive. People won't work together so much anymore. They won't know each other as well. They'll lose their connections. And the community itself will begin to come unglued.

Interestingly, this all goes back to that first letter I received from the State Engineer, in the summer of 1969, that offer of adjudication of water rights on the Pacheco Ditch for my acre and a half of land. Of course, I had no idea what they were talking about, but pretty soon I would learn. The adjudication was, and still is, an attempt to legally define every drop of water in the Upper Río Grande watershed. It was launched up in Taos not long after Aamodt was started down here in the Nambe/Pojoaque area.

Concurrent with the adjudication, the state was also trying to partition San Juan-Chama Diversion water to various areas of New Mexico, including Taos. For our valley, the plan was to give us 12,000 acre-feet of water to be impounded in a dam called the Indian Camp Dam, that would be built just south of Taos. In order to build the dam and contract with the Bureau of Reclamation and the state we were told that we had to form a Conservancy District in order to tax the people in the Taos Valley for all the good fortune slated to come our way as a result of the dam.

At first, everyone was for the Indian Camp Dam, because who do you know in New Mexico who isn't for more water? In a state this dry you'd have to be crazy to say "NO" to more flow. But pretty soon a lot of folks realized that even if the dam was subsidized in part by the government, the cost to locals would still be pretty steep, given that this is a real poor area. But after the realization people really rebelled when they began looking into the legal powers of a conservancy district, and they concluded that in all likelihood the control of water in this valley would pass from the individual communities and acequia systems to politically appointed boards who would very likely shift much of the water in this area into developmental endeavors. And, given the nature of development endeavors in New Mexico during the Twentieth Century, the local farmers and ranchers and other residents figured they were going to get screwed.

So they formed an organization called the Tres Ríos Association to fight against the Indian Camp Dam, the Bureau of Reclamation, the State Engineer's Office, the Conservancy District, and many of the bankers and lawyers and developers and business people in the Taos Valley. The Tres Ríos Association was made up of almost all the acequias in the valley, and most of the people on those acequias. The battle lasted for a good part of nine years, and it created a fair amount of bad blood in the Taos Valley. It pretty much pitted the development future against the sustainable past, and it certainly put the history, and custom, and culture of the Taos area up for grabs. The San Juan-Chama water slated for Indian Camp Dam was tied into the adjudication of all the water rights in the valley.

It is not an accident that most of the leaders of the Tres Ríos Association were elderly gentlemen and women and many of the meetings were conducted in Spanish. Many of those leaders are gone today, and the Taos Valley—and myself—misses them dearly. They were not radicals, they were not people who wanted to ask for trouble, nor were they ostriches who decided to stick their heads in the sand in order to avoid facing the realities of modern growth and change. They were people deeply rooted in a culture and history that had shaped New Mexico for centuries. They were republicans and democrats, they were farmers and teachers, they were veterans and pacifists. Some had herded sheep, others had worked at Los Alamos. They were grade school principals and people who worked on the county roads and men who built houses and women who ran little grocery stores.

The oral history of the Taos Valley was repeated, explained, and venerated at every meeting. All the feuds in the valley could be present in the background at any meeting. All the politics of the valley were present at every meeting and heatedly debated by the various participants. At the meetings were people who loved each other, and people who hated each other, but the really important thing was that the people dealt with each other as a community. That is the way democracy is supposed to work, with all the participants personally involved.

There was a time during the heat of the battle against the conservancy district in Taos that I had tacked up, on every wall in my little adobe house, the State Engineer's hydrographic survey maps of every irrigated piece of land in the Taos Valley. I knew by heart all the acequias, their locations, and many of the pieces of land that they irrigated. I had a big telephone-book-sized list of all the parciantes. I knew many of those parciantes personally. It was the most intimate kind of map you could have of my home area. When I looked at all those parcels of irrigated land and the people who owned them I was learning an entire town. It was like being in medical school and dissecting a body. It was like memorizing the Bible. It was like learning the entire history of a people that had become an important part of my own history and vice-versa. It's one way I put down roots. That's how communities remain strong, when their people have that connection, those roots, that obligation.

The Tres Ríos Association raised a lot of money to pay for lawyers. The leaders went back and forth to Santa Fe and back and forth to Santa Fe and back and forth to Santa Fe. The old timers' tenacity amazed me. They hired people to do research and then they listened to the result of the research. They attended countless court hearings. Sometimes nobody came to a meeting, so the leaders called another meeting, and nobody came to it, so they called a third meeting. . . . and all of a sudden everybody came.

Eventually, despite our protests, the district court formed the Taos Conservancy District anyway. So we appealed the verdict and kept fighting. And a few years later the state Supreme Court overturned that district court decision. And in the end the conservancy district and the Indian Camp Dam did not arrive in Taos. But the adjudication never quit. And, of course, development has gone on anyway, although it received a serious setback when Taos shot down the conservancy.

Ever since, people have been struggling to maintain the traditional acequias and community water structures and all they represent. The adjudication suit is a difficult and cantankerous beast. It is an attempt to once and for all impose an American legal system on an area that has operated on largely Spanish custom for the last few centuries. The adjudication wants to clear things up so that water can be bought and sold with impunity, clearing the way for modern growth and development. It wants to make it possible to evolve from so-called "inefficient" and antiquated ways into the fast-paced 21$^{st}$ century.

In the process, the adjudication suit has pitted people against each other, and one acequia against another acequia in a scramble for more advantageous priority dates. Indigenous Latino irrigators are placed against irrigators from the Pueblo, which entered the adjudication asking for enormous amounts of water. The State, the Feds, and a passel of lawyers seem to have grown fat over this. In Taos, the old Tres Ríos Association segued into the Taos Valley Acequia Association that continues to represent the often impoverished parciantes on most of the ditches in the Taos valley. Many people still don't understand this convoluted process.

Now: Ever since I came to New Mexico in 1969 I have read articles in the newspapers about Aamodt vs. the State of New Mexico. It was your 800-pound gorilla down here like the Indian Camp Dam and the conservancy district were Taos's 800-pound gorilla up there. I never really understood what was happening with Aamodt, the same way that for years nobody really understood what was happening with the adjudication, the Indian Camp Dam, the conservancy district, and the San Juan-Chama diversion water in Taos. I don't know what your experience has been down here, but up in Taos years ago, official representatives would arrive from the State Engineer's office—like Paul Bloom, or Eluid Martínez (who eventually became the state engineer)—and they would call meetings to explain to us the conservancy district, the proposed dam, the San Juan-Chama Diversion Project, the adjudication suit, and also Einstein's Theory of Relativity. And after Paul Bloom and Eluid Martínez had spent five hours explaining all these things in both Spanish and English to maybe 100 small farmers and teachers and construction workers and cabinet-makers sitting in the auditorium of the García Middle School, there would be a long pause. And then one of those small farmers would stand up and say, "That's all very well, pero no lo hacemos así aquí en Taos—but we don't do things that way here in Taos."

And after a while the government authorities and the Bureau of

Reclamation and the state and federal tiburones (I mean abogados) and local development sharks realized that they were talking to a wall.

And the wall never collapsed.

One reason the wall did not collapse is that the people of Taos realized the adjudication, the San Juan–Chama Diversion, the Indian Camp Dam and the conservancy district were probably going to ream them completely. Individual community acequias were going to lose their autonomy to a politically appointed conservancy board, i.e., we were going to lose control of our water systems. And we were going to have imposed upon us an open-ended taxation system over which we had no control. We weren't even going to be allowed to vote for the conservancy board that would govern us because the powers that be recognized that if we were given the opportunity to vote we would immediately dissolve the conservancy and that would end the Indian Camp Dam project. The Tres Ríos Association studied the history of other major water projects along the Río Grande, especially the Elephant Butte Dam and the Middle Río Grande Conservancy District, and it discovered that in every case that water projects were advertised as ways to help local farmers grow more and better crops for economic profit, they wound up instead running indigenous people off the land in favor of agribusiness corporations and urban/suburban development.

A fascinating study of this process was done way back in 1936 by a Middle Río Grande Regional Conservator, Hugh Calkins, in a document called "A Reconnaissance Survey of Human Dependency on Resources in the Río Grande Watershed." Calkins explained that before Elephant Butte Dam 70 percent of 889 farms in the area were "owned by their operators," a large majority of whom were "Spanish-Americans." He described the farms and people as self-sufficient and relatively stable. Then the dam and irrigation district arrived and suddenly bankruptcy and loss of farms "became a constant threat." And in the end, "The irrigation project . . . was the instrument by which this essentially self-sufficing area was opened to commercial exploitation. The establishment . . . of a legal claim upon the resources of the area and the labor of its inhabitants . . . led to the dispossession of the natives, and their replacement by American settlers financed by American capital . . . Through the construction of a costly irrigation project . . . an additional land area of 100,000 acres was made available for agricultural use," but "the native population, unable to meet the new high cash costs, was in large measure displaced from the 50,000 acres it had owned."

A noted sociologist, Dr. Clark Knowlton, who testified at several 1970s conservancy hearings in Taos, had previously written: "Every major irrigation or water conservation project along the Río Grande River . . . has been responsible for land alienation on an extensive scale. The Spanish-Americans have been replaced by Anglo-American farmers. Their subsistence agriculture has made way for a highly commercial, partially subsidized, and basically insecure agriculture, made possible by government programs. Little thought has ever been given to the rights and land use patterns of the Spanish-Americans in planning water projects in New Mexico and neighboring states."

Needless to say, little or no thought has been given to the rights and land use patterns of the Native Americans during this same epoch, until recently. And even then it often seems that today Pueblo rights are merely being touted as a way to partition and confuse indigenous communities in a form of divide and conquer, weakening both sides, so that eventually most of the water will end up in golf courses for rich tourists, luxury hotels, developments like Las Campanas of Santa Fe, and in frantic urban development from which native and Latino and all working-class New Mexicans will be excluded.

The proposed settlement agreement and regional water system being offered to the people of the Pojoaque/Tesuque/Nambe area reads to me like a nightmare of antagonistic and far-fetched water development. For starters, all the water proposed for future use in the area only exists on paper. It will cost a fortune to develop a regional system, but nobody can tell if the water really exists for this system or how much it will actually cost. I read in newspapers that the government will pay "most" of the estimated 280 million dollars for the system. During the conservancy and dam battle in Taos we were told the government would pay something like 96.5 percent of the costs, but we figured out that that last 3.5 percent plus maintenance of the dam would not only be exorbitant for the relatively poor population, but it was open-ended with no guarantee there wouldn't be endless cost overruns.

Today in the Southwest we are locked in a severe drought. Global warming is a fact of existence. In the last few years I have read 100 articles by scientists, sociologists, hydrologists, political and demographic analysts declaring that most of the major wars of the 21$^{st}$ century will be fought over water. It's not just in New Mexico, it's everywhere. U.S. America's Oglala aquifer is draining way down; the Aral Sea in Uzbekistan has half evaporated and is a disaster area. And you know that Israel will never give Palestine its own state because something like 70 percent of Israel's water comes from the West Bank.

I think if I had current irrigation rights or access to well water I would not in a million years give them up for paper water projected to cost me a fortune with no guarantee of future delivery in the first place. At the rate we're going, the state of New Mexico could very well sink 280 million dollars into a regional well system that'll just wind up drawing sand from the Río Grande. The whole gambit is a rush to more urban development in a desert state already buggered by nonsensical growth which seems like a formula for Land of Enchantment suicide. History teaches us that for sure the regional water system project is not geared to benefit the working men or women of this valley, or folks of modest means. It is more like a grand ploy to stimulate outrageous growth-oriented commercial development for the rich, and as such just another criminal boondoggle in a global world economy eager to self-destruct. I apologize to Indian water users who have been short-changed for centuries by the outside world, but we are now living in times that cannot tolerate more development insanity like golf courses and luxury hotels in a desert which ultimately will destroy the biology that nurtures all of us.

Do you remember the 80s under Ronald Reagan when the savings and loan industry collapsed, much of it because huge loans in the oil and gas industry were being given based on a collateral of future production that never materialized when the energy industry tanked? This disaster cost American taxpayers untold billions. Well, even as I speak, given the drought, global warming, and expanding human population, future water, as collateral for proposed regional systems and urban development, is becoming a pipe dream. Edward Abbey once said, "Growth for the sake of growth is the ideology of the cancer cell." We should be stopping all growth, going on emergency water rations, and completely retooling the economic philosophy that guides our behavior. Instead, we keep frantically trying to tread water and expand our numbers in a lake that increasingly is running out of $H_2O$.

A major serious problem confronting all of us is philosophical. It's all about ideology. It's about our attitude toward our own lives of material well being. If we need to keep consuming as we do, if we need to keep driving gas-guzzling SUVs as we do, if we need to live in big houses with air-conditioning as some of us do, if we need to have lots of gim-cracks and gee-gaws and entertainment centers and vacations in Mexico or Hawaii . . . if, in short, we need to be typical American consumers in a world where our habits are causing 27,000 species to go extinct each year and a forest the size of New York State to be logged each year and almost three billion people elsewhere on the globe live on less than two dollars a day . . . then there's no point to resisting the proposed regional water system to settle the Aamodt case. Because as far as I can tell the settlement is an attempt to keep up U.S. America's excessive material lifestyle against all evidence that the lifestyle is destroying the planet we live on.

Resisting the settlement should indicate a desire to call a halt to the growth and the consumption that threatens life on earth. But so far not many U.S. Americans have been willing to change our lifestyles to bring about a more sustainable order.

So my feeling is this:

The development future being offered to us in New Mexico and around the world is grim indeed. Here in Northern New Mexico we still have partial access to a precious way of life. From my outsider's perspective, the current solutions offered to Aamodt seem negative for all parties concerned. If possible, I would band everyone together in this valley to demand a more benevolent social, economic, and environmental solution that does not destroy the infrastructure that took hundreds of years to produce by crushing it under a massive water project that has enough costs and other hidden variables in it to sink a battleship. Authorities have told you that if you don't accept the current settlement, you'll lose everything: I'd guess that is BS. Thirty years ago Helen Ingram wrote a book called "Politics in Water Resource Development" in which she explained how water projects are sold to communities. I quote:

> "Support for a water project is justified by magnifying the need and benefits of water projects. It is claimed that projects will foster all sorts of goals, grandiose or particular, even if these goals may be contradictory ... Economic and social advantage is promised to all sorts of groups, even if their interests conflict ... Crisis in terms of water scarcity or floods is exploited to create consent. Backers of projects claim that an emergency situation exists. Projects are said to be essential to economic survival. At the very least, it is asserted that continued growth of a locality hinges upon the particular development being pushed ... Projects are made to pass experts' tests even if there is little agreement among the experts on the soundness of these tests ... "

And so forth.

The fact is: If many people in this area believe the current solutions to Aamodt are unjust, unfair, and untenable, the fight ought to go on until a more equitable, and a more sane conclusion is realized.

You know, it's been a while since I've walked the length of an acequia in Taos. Today, I live in a small house, I have no irrigated land, I'm not a commissioner any more. But I think often of the Pacheco and the Lovatos ditches that used to give me water. I can see myself walking along the bank with a shovel, checking it out, looking for problems. I walk through yards and little back fields. There's Lucy Mares' house and her daughter Stella's trailer. And Vidal Cisneros' big purple martin birdhouse, and Shorty's Mower Service garage. Indian ponies on Pueblo land shy away from me, they never come over for handouts. Peacocks are strutting by Isabel Vigil's corrals, and her little donkey starts braying. Tom Trujillo's old horse is standing knee deep in grass in his daughter Frances' front field: And they're singing gospel hymns in the Good News church that's right behind Sebastian's old bar. A rusted truck is sinking on its axles in Adolfo Lavadie's two acre plot. And somebody's hanging out wash by Archie Anglada's trailer. One of the Pacheco boys is shooting at a prairie dog in their garden: "Orale, bro', cuidadito!" And we're gonna have to chop down all these jaras behind Miera's field also. And I better call the Córdovas and tell them it looks like there's a hole in the fence where the sheep could move through to the Romero's pasture where the alfalfa would bloat the idiot sheep to death in fifteen minutes. And when can we get a crew to repair the desague just below the Medina's corrals?

Water can still be like that in Taos, in Northern New Mexico: up close and personal. Nobody is making much money from irrigating small pastures and little gardens of acequias and old wells drilled long ago. People don't hang onto the acequias or little wells and fight for them because of all the profit involved anymore. No: We hang onto this way of life and fight for it because water and the local organizations that dispense it are the blood that keeps our communities alive.

In Taos, people say this: "Buen abogado, mal vecino."

They also say: "Sin agua, no hay vida."

When Reies Tijerina was up north fighting for the land grants I would see signs everywhere that said: "Tierra o muerte."

And I myself always end every talk I give by saying: "Hasta la victoria, siempre!"

## *10*

# The Writing and Filming of The Milagro Beanfield War

### John Nichols

*The following passages are excerpted from the Introduction to A Fragile Beauty (1987) by permission of the author.*

The Milagro Beanfield War was the first book I published while living in New Mexico. As such, it broke ground for my other novels like *The Magic Journey* and *The Nirvana Blues*, and for the non-fiction book: *If Mountains Die* (with photographs by William Davis), *The Last Beautiful Days of Autumn*, and *On the Mesa*. Different as the emphasis of each of these books may be, they all deal with questions of land, cultural ethics, problems of ecology, economics, history, and human survival.

John Muir once said, "When we try to pick out anything by itself, we find it connected to everything else in the universe." I have always kept that in mind while writing about the land, people, heartaches of northern New Mexico. To extol the fragile beauty of the Taos Valley in words, photographs, or in a film, is to sing the praises of, and to demand consideration for, the entire earth.

The implications are always universal.

It's an old adage that the origins of things reveal their meanings. My love of the Southwest began in the summer of 1957, when I first visited New Mexico and Arizona. I was sixteen. I spent a week in Taos plastering a friend's adobe house, then headed down to a New York Museum of Natural history research station in Portal, Arizona, not far from the Mexican border. There I spent some time as a ranch hand, as a carpenter, and I helped scientists collect bugs, beetles, butterflies. I also fought a few forest fires as a volunteer smoke chaser, earning $1.50 around the clock.

It was a dream summer in a dream territory.

When, in 1969, I left New York, New Mexico was a logical place to head for. Not because of land or majestic weather. Rather, New Mexico seemed to resemble a colonial country where political struggle could be as clearly focused as it was in four-fifths of the rest of the world. I had been reading a newspaper called *El Grito Del Norte*, published by Betita Martínez in Española, near Santa Fe. Copies showed up regularly at the 8th Street Bookstore in Manhattan. I also read Stan Steiner's books about the Indian and Chicano movements, *The New Indians* and *La Raza*. I was fascinated by Reies Tijerina's land grant politics which had culminated in an armed raid on the Tierra Amarilla, New Mexico, courthouse in 1967.

Yet in the end, in New Mexico, it was the land, as much as the social and economic situation, which fired up my energy and gifted a new life I had not dreamed of. I hiked the wide sagebrush mesa west of Taos, gathered pinon wood in the hills made redolent by autumn wildflowers, worked on irrigation ditches that carried water to gardens I began to grow. And I soon grew to love fishing in the prehistoric Río Grande Gorge; I was even captivated by the cruel, alarming iciness of winter weather at 7,000 feet, in the shadow of 13,000 foot mountains.

The land was essential to the rhythm of life in Taos, and all of it was threatened by one form of development or another. Sadly, I quickly came to realize, the longtime caretakers of the valley were being driven out.

Shortly after reaching Taos, I began writing for a muckraking journal called *The New Mexico Review*. Originally founded by a fellow named Ed Schwartz, who considered the paper a legislative watchdog, it was soon taken over by a Princeton hockey player and a fledgling lawyer named Em Hall, and his buddy, Jim Bensfield. They couldn't even pay salaries. Indeed, I doubt they ever even paid for a story. To make up for this shortcoming—at least in my case—Em Hall played center on a line with me in an Albuquerque ice hockey league, and he continuously fed me the puck so that I could score goals. Such unselfishness from Em, given his own savage lust to rack up points, touched me more than I would have been moved if he'd simply paid me for my writing, but had then refused to pass the puck!

The *Review* ran on a shoestring and a prayer, a bit of advertising revenue, and small grants and donations from the lunatic political fringe. Hall and Bensfield would have liked it to be a southwestern *New Yorker*, with a bit of aristocratic trout fishing thrown in. Unfortunately, just to keep the paper going, they were forced to let a collective run it. And much of the collective—including yours truly—had more radical dreams for the rag. Of course, the farther left we went, the harder it was to dislodge bread, even from the liberals. And by 1972 the *Review* was going down fast. Bensfield and Hall turned over the editing to Jim Rowen, George McGovern's son-in-law. But Rowen grew disillusioned after a few issues and decided to bag the venture. I couldn't bear to watch it die, because I loved it, and also, I suppose, because the *Review* was the only organ to publish me since 1966!

So I became volunteer editor for the last three issues in 1972. A Taos artist, Rini Templeton—who would later illustrate *The Milagro Beanfield War*—taught me about makeup, layout, theories of design; together we kept the magazine going.

But by November I was photographing, writing, and cartooning almost the entire issue. And, exhausted, dismayed, I gave up; the *Review* finally expired, never to be resurrected again except as *The Voice of the People* in *Milagro*.

For those three years the magazine provided a forum for my views. It also gave me an excuse to research the valley in which I lived. My first article concerned a friend, Joe Cisneros, who had been fired by a molybdenum mine in Questa, a town half an hour north of Taos. My style, at that time, was not exactly laid back and "objective":

> *In short, Joe Cisneros has committed the unpardonable sin of confronting Corporate America. Almost single-handedly he has locked himself in combat with the monolith ... Joe was fired for trying to convince his fellow workers at the mine that they should stick up for their rights; he was fired for protesting the hiring, firing, promotion, job assignment, and disciplinary policies at the mine; he was fired for standing up to the Anglo bosses who traditionally have looked upon the Chicano much as the southern plantation big shots have looked upon the Negro. And finally, Joe was fired for his activities outside the mine, where he dared to stand up to the Corporate Giant, saying, "Your greed stops here, you don't get my land, I've had enough."*
>
> *And when you read the facts of how Manifestly-Destined Anglo America has systematically stolen the land in this part of the country, presuming it has had the God-given right to do the peasants out of everything but their underwear for the past handful of centuries, you begin to realize how these defiant gestures by Joe Cisneros might rankle.*
>
> *So they fired Joe.*

That article was published in May 1970. Though I didn't realize it at the time, Joe Cisneros was a prototype for the kind of person who would become Joe Mondragón, chief protagonist in *The Milagro Beanfield War*. My article dwelled at length on Joe's imaginative struggle to get by. He made hunting knives out of used Skil saw blades and sold them for six bucks a pop; he had fashioned an air compressor out of an old refrigerator motor and a junked sander; he had fabricated a new tractor muffler from a piece of pipe drilled full of holes:

*And listening to Joe describe his projects and explain how things work, you understand how necessary it is, if you are poor, to learn how to squeeze blood out of a stone. In order to survive you must be able to coddle a machine with spit and baling wire and whatever else is at hand, because often the machine absolutely must function in order for you to eat; and with a little luck and hard work and ingenuity, what other people throw away, the junk in Chicano yards that so offends the tourist from Boston, can keep you going indefinitely.*

*So Joe is not only a highly skilled workman, he is also a man who can work miracles, coaxing an extra four or five lifetimes out of things built to fall apart after two years. And thus deep down, at a most intimate level, Joe is the system's enemy.*

Shortly after the article on Joe Cisneros appeared in the *Review*, I drove to Costilla, a small town on the Colorado border an hour north of Taos. At the time, Costilla was very nearly a ghost town. West of the main highway it was composed largely of ruined old adobes crumbling back into the earth, barren fields, blown-over outhouses, collapsed and splintered roof timbers. East of the highway, pastures were green and irrigated. I spent but a single afternoon in Costilla; yet the ruins made a lasting impression on me. When I asked a resident about the desolation, she replied vaguely, "Oh, you know, the state took it away. Some people, they had a kind of association, some pull . . . and took all the water. They did it to the poor people . . . it's the poor people who lost the water." Apparently, most of the ruined houses had belonged to bean farmers who departed when they lost their irrigation rights. She smiled: "This used to be a very beautiful place . . . ."

I wrote an article about Costilla, the lost water, the crumbling ghost town. The village became a metaphor that remained in my head. And, although I knew nobody in that town, and did not return to it for years, fictitious Milagro had been born.

Almost every month I contributed a story or two to the *Review*. I wrote about a Chicano-hippy war in Taos, the history of Los Alamos, police brutality. Kit Carson—the patron saint of commercial, tourist Taos (who is often seen by Native Americans as an agent of U.S. genocide)—fell victim to my outraged pen. I supported the Taos pueblo's struggle to recapture its sacred Blue Lake land from the Forest Service. I eulogized Chicano activists murdered by the police, publicized feminist artists, supported a liberal Taos high school social studies course that was derailed by angry parents who denounced it as "socialist." I covered a speech by Reies Tijerina, and documented a visit to New Mexico by Luis Valdez's Teatro Campesino. The Teatro was a political theater group whose raucous, comic style I would eventually try to incorporate into *The Milagro Beanfield War*.

During this time I also did support work for various political groups. One was a cooperative and clinic in Tierra Amarilla, two hours west of me. Another was a Taos organization called Trabajadores de la Raza. I touched down occasionally with a Santa Fe barrio group called La Gente. I remember one fearful night riding police brutality patrols with them. Another time, at a Colorado hospital auction, I purchased beds, lamps, and other supplies for a people's clinic opened by La Gente. Often, I went to Española to help in mailings of *El Grito del Norte*. Too, I chronicled water misuse by recreation developers in the Valle Escondido eight miles up Taos Canyon. And I repeatedly criticized the Forest Service grazing and woodcutting policies, which were prejudiced against many longtime residents of the Taos Valley.

One day I drove a carload of old folks from Peñasco down to the Santa Fe legislature for a demonstration against proposed welfare cuts. On the journey south, I chatted with an old man who recounted a wonderful story about his father. I published his tale in the May 1971 *Review*:

> *My father now, he's 96 years old, and I got nine brothers and sisters, all of us living. When my father was 79, there was a big operation to save his life. We all thought he was gonna die. The doctors said he had at most six months to live without the operation, maybe two years or so if the operation was successful. As a family we had to vote whether the operation was to be done or not. One of my sisters was sure the operation would kill him, and voted no. But anyway, they did the operation, and although he was pretty bad for a while, he survived. That Christmas my father called us all together to have a last celebration, because he didn't think he was gonna live much longer. So we all came to be with him*

*on his final Christmas. My brothers came from California, even, and everybody said goodbye. Then, when he was 84, the old man wanted us all to gather again, because it was gonna be his last Christmas. Well, not everybody came because not everybody believed him. Later on, when he was around 90, or something, he wanted us all to come again, but a lot of us were too busy, and besides, by then who believed him anyway? He was real sickly when he was a child, and he's been sick a lot his whole life. Just a couple of years ago he was operated on again, and I was sure when I looked at him in that hospital that he wasn't going to walk out of there. But he did, and he can still read the newspapers. . . .*

In *Milagro* that story became the foundation for an indestructible character, Amarante Cordova, whom I saw also as a metaphor for a culture that refused to die.

In fact, it was the old-timers around Taos who most clearly conveyed to me a sense of the land, the agricultural rhythm of life, the culture, politics, and economics of the valley. Granted, I worked with many people my own age on various projects—organizers like Kelly Lovato, Jerry Ortiz y Pino; progressive lawyers like Peggy Nelson and Tony López, or like the Taos enjarradora, Anita Rodríguez. I visited Tierra Amarilla a few times in support of clinic functions put together by folks like María Varela and Moises Morales and Dr. Ed Bernstein, or La Raza Unida members Ike De Vargas and his hardworking mouthpiece, Richard Rosenstock. I attended benefits in support of the Ensenada Velásquez family, which was defending its land against an attempted takeover by a Río Arriba ski area developer. I wrote an article on their principle spokesman, Esteban Polaco, whose determination to protect his family's roots had a lot in common with my friend Joe Cisneros—and the fictional Joe Mondragón.

But the old-timers most powerfully gifted to me the Taos Valley. Some, like my immediate neighbors, Eloy Pacheco and Bernardo Trujillo, taught me the lore and structure of the two irrigation ditches running through my property. Another, Melitón Trujillo, who lived across the river from my house, talked about old-time politics and sang corridos he had invented during Republican-Democrat frays of old. Jim Suazo, at the Taos pueblo, was a dignified and powerful storyteller; his son-in-law, George Track, was a loquacious Sioux adventurer whose extended family and convoluted, cliff-hanging lifestyle always touched me deeply.

Andrés Martínez entered my life in 1970, when Taos was slated for a conservancy district and a large water impoundment project to be made possible by construction of the Indian Camp Dam. Andrés presided over a coalition of valley acequias against the dam, called the Tres Ríos Association. Older men allied with Andrés, who also became my friends, were Paul Valerio, Bernabé Chavez, J.J. García, Jacob Bernal, Pacomio Mondragón.

For years I worked with the Tres Ríos Association while it fought the conservancy and the dam. People protested because the socioeconomic changes to be triggered by the enormous development project would threaten the valley's marginal citizens—mostly Spanish-speaking small farmers unable (and unwilling) to compete with high-pressure capitalist ventures.

Tres Ríos meetings were the best education I could have gotten about the social structure, personality, and history of Taos. I also did research for Legal Aid lawyers working on the case. In the process, I learned more about land and water rights history than I could ever have dreamed possible. I also drew cartoons for newspaper ads against the dam, and designed flyers to protest it or to call the public to the meetings. Upon occasion I hiked from one end of Taos to the other, leafleting automobiles with yet another anti-conservancy call to arms. At one point I had the State Engineer's hydrographic survey maps of the entire Taos valley pinned to the walls of my kitchen, living room, and bedroom as part of an effort to learn the landholding patterns of the entire valley. I spent hours in the county courthouse researching titles, attempting to discover the true ownership behind dummy corporations.

And I wrote about the Indian Camp Dam controversy for the *New Mexico Review*. In the July 1972 issue, I summed up the problem like so:

> *The fact is, the history of Conservancy disasters in New Mexico, in which the small Chicano farmer has always been the fall guy, is lengthy and depressing. It can be summed up in these words from noted sociologist Dr. Clark Knowlton:* "Every major irrigation or water conservation project along the Río Grande River, from Elephant Butte Dam to the Middle Río Grande Conservancy district, has been responsible for land alienation on an extensive scale. The Spanish-Americans have been replaced by Anglo-American farmers. Their subsistence agriculture has made way for a highly commercial, partially subsidized, and basically insecure agriculture made possible by government programs. Little thought has ever been given to the rights and land use patterns of the Spanish-Americans in planning water projects in New Mexico and in neighboring States."

It was a classic battle of the big boys versus the little guys. Allied against the Taos small farmers were the State Engineer, Steve Reynolds, and his right-hand gunslinger, Paul Bloom (whom I repeatedly jabbed at with *Review* cartoons). The Bureau of Reclamation was eager to build the dam. Most realtors, bankers, lawyers, motel and hotel owners of Taos favored the project. All the money was stacked on the side of progress, American style. Yet Andrés Martínez and his cohorts, who began with almost no precedent, no money, and no political power on their side, ultimately defeated the conservancy district and the Indian Camp Dam.

To me, the Taos conservancy struggle was a lot like the 1950s film *Salt of the Earth*, which I first saw at Em Hall's house in Pecos. The movie is about a successful miners' strike in southern New Mexico. It not only treats Chicanos and working class people with respect, but it is also a powerful feminist film, and was long banned or suppressed in this country because of its "communist sympathies." Later, when I wrote *The Milagro Beanfield War*, I hoped that I could emulate successfully the message and humanity of that film.

So I discovered the history of Taos through Andrés Martínez and many other friends. I learned about community complexities, the feuds, the intricate and extended family ties, the political shenanigans. Most importantly, I absorbed my neighbors' ties to the lands, and participated in their—in our—struggle to defend it.

Meantime, as the *Review* floundered towards its November 1972 extinction, I did some personal floundering of my own. I had published *The Sterile Cuckoo* in 1965, *The Wizard of Loneliness* in 1966. After that, my literature became so infused with political rage that I hadn't published a thing. I was thirty-two and money was scarce—almost nonexistent. My marriage was on the rocks. The mortgage on my four-room adobe house was only $78.38 a month, but I often had trouble making the payments. I heated with wood, used an outhouse, and didn't even have a car. It looked like curtains for the not-so-young-anymore writer whom H. Allen Smith once suggested in print "will likely become the best comic novel writer of his time . . . ."

Truth to tell, I didn't feel very funny anymore. I was torn between trying to write polemical books, or just ditching my artistic pretensions and gravitating toward more immediate political barricades.

In the end, I chose to give political literature a final shot. What followed has become slightly apocryphal—at least in my eyes. During one of our hellacious hockey matches in Albuquerque, I told Em Hall the *Review* was dead, I could not afford to edit it any longer. He passed the puck anyway, and I scored a few more goals. Then I sat down and began a new book on a one-sentence premise: a 35-year-old unemployed Chicano handyman named José Mondragón (who was about as desperate as myself) cuts water illegally into a half-acre beanfield, and all hell breaks loose. Eventually, his act goads a community of impoverished, working-class people to organize against a big dam and recreation development (made possible by a conservancy district) that threatens to wipe them out.

The kitchen was freezing, so I typed fast on my little Hermes Rocket to keep warm. I churned out the first 500-page draft of *The Milagro Beanfield War* in five weeks. I took three more weeks to correct it, another three weeks retyping the book. At that point the Hermes collapsed, I chucked it into the trash, and sent the book off to my agent, Perry Knowlton at Curtis Brown in New York. By then it was the end of January 1973. Perry gave the book to Marian Wood, an editor at Holt, Rinehart & Winston, and Marian promptly bought the book for 10Gs. I was astonished. Only four months had passed from the start to the score.

After the long famine such a sudden feast!

Then I scrambled to rewrite the book. Holt gave me eight more months to "tinker." While I was tinkering, producer-actor Tony Bill dropped by Curtis Brown and asked, "Anything new from John Nichols?" (Tony is a guardian angel who gave my first novel, *The Sterile Cuckoo*, to Liza Minnelli, who starred in Alan Pakula's film version, which kept me passably alive from 1966 until I came up empty in 1972.) Perry handed Tony a second draft *Milagro* manuscript, and Tony gave the manuscript to his friends Bob Christiansen and Rick Rosenberg, producers of "The Autobiography of Miss Jane Pittman." They liked it, got in touch with the Curtis Brown Film agent, Richard Parks, and all of a sudden *Milagro* was out on option eight months before publication! Plans were to hire Mark Medoff, of "When You Comin Back Red Ryder?" fame, to do the script. Tomorrow Entertainment would spring for the development bread. Boy, what a reversal of my fortunes!

Of course, Hollywood has long been a reactionary industry which generally has treated minority people in a patronizing or downright racist manner. A good reason, I initially surmised, never to let *Milagro* out on option. Aside from rare films like *Salt of the Earth*, *Zoot Suit*, or *The Ballad of Gregorio Cortéz*, the traditional image of Latinos has been either as thugs, as the butts of dope and toilet jokes, or as quasi-morons obsessed with gang warfare and hydraulics.

Nevertheless, as a film freak from age seven, I have always understood that good films *do* get made. And so my idealist side believed a *Milagro* film could break the worst Hollywood stereotypes. Still, it was a risky adventure. If the outcome was patronizing, all option and other payments I'd taken would seem like blood money in the worst sense. But if the film had a political and cultural integrity, then, hopefully, it would count in the struggle of all people to assert their basic human rights.

Unfortunately, after writing an outline of *Milagro*, Mark Medoff disappeared into the convolutions of his theater world and proceeded no further on the script. Chris-Rose left Tomorrow Entertainment. Meantime, *Milagro* was published to fairly good reviews and a semi-resounding commercial clunk. Holt did a small second printing, but then abruptly remaindered the book.

I was broke again, but I invested $200 to buy some 400 hardback copies of the novel, and I passed them out like canapés to any friend, foe, fan, or local bookstore that was interested. I thought I'd have hardback copies for the rest of my life; instead, within months all my originals were gone.

At the last minute, paperback rights sold to Ballantine for $7,500—so, financially, I was saved by the bell. Fresh off that triumph, for $40 I bought another Hermes Rocket—the world's last, great disposable typewriter—and churned out two more novels, both of which were immediately rejected. And very shortly I was right back where I had started in 1972.

To Bob Christiansen and Rick Rosenberg, who kept renewing the *Milagro* option, I owed my survival. By 1975 Paramount was fronting the development bread. Chris-Rose hired Tracy Keenan Wynn, the author of "Miss Jane Pittman," to write the *Milagro* script, and I was ecstatic. What could go wrong now?

Well, time passed quickly—1976, 1977. Chris-Rose persistently renewed the *Milagro* option, which kept the project *and* me alive. My literary career was dead again; I was struggling with a novel called *The Magic Journey*. While I stumbled toward completion of that new book, *Variety* announced that Paramount would soon be filming *Milagro*. But Tracy's script of the novel had problems. Biggest one, no doubt, was trying to distill a 600-page novel with some 200 characters down into a two-hour "ensemble" piece, in which at least half the characters spoke Spanish as their primary language.

Though I'm a novelist at heart, I've long believed the axiom that to make a film, "you buy the book and throw it away." Here, however, we were also dealing with executives whose initial reaction to the project was, "Who wants to see a movie about a bunch of Mexicans?" Particularly a bunch of Mexicans who took on the traditional Anglo power structure, winning a victory that went completely against the grain of progress, American style. Too, the moneymen wanted a "star" attached to the movie. A Chicano star? Not at all. They were talking Al Pacino, Dustin Hoffman, Robert de Niro: I cringed.

So the project never really caught fire, Tracy's script receded into the background, Lorimar stepped in to take over the development financing, and Leonard Gardner, of *Fat City* fame, tackled a new screenplay. *Hollywood Reporter* announced that Lorimar would soon start filming *Milagro*. Warren Bayless, my new film agent at Curtis Brown, advised me not to hold my breath.

During this time, on the home front, the Indian Camp Dam battle continued. A district judge formed the conservancy district in Taos. The Tres Ríos Association appealed that decision to the state supreme court. And, on a technicality, the supreme court overturned the district decision, throwing out the conservancy district. I'll wager that this was one of the few times in modern southwestern history, that a people as poor and politically "powerless" as the Taos farmers had managed to organize successfully against a state—and federally—supported irrigation project of such magnitude.

Life imitates the art that imitated life!

Regarding the movie, Leonard Gardner's *Milagro* script wound up lacking energy. It was rewritten umpteen times; yet despite the best of intentions it eventually became obvious that this effort was also doomed. *The Magic Journey* was published, went nowhere. And, as the winter of 1978 approached, I was broke and cold once more, desperately flailing away on a new novel, *A Ghost in the Music*, and a non-fiction book, *If Mountains Die*.

When it was published, *If Mountains Die* featured beautiful color photographs of the Taos Valley by my friend Bill Davis. The book is gentle compared to *Magic Journey*. Yet at many levels—and particularly because of Bill's evocative photographs—I felt that it made a strong political statement. Though no slogans clutter Bill's serene landscapes, I learned from the success of our book that beauty itself is an effective weapon in the battle to revere and protect life.

The year *If Mountains Die* came out (1979), I began packing a camera on all excursions about my home territory. In the process, I soon realized that all landscape (and photographs of it) is defined by human attitudes and sensibilities. In particular, all Taos scenes I witnessed seemed powerfully connected to the legacies of cultures which have dominated the valley for centuries. Because nature is essentially at our mercy, each beautiful landscape I photographed seemed to praise the human history of this place.

When the *Milagro* option expired in 1979, Bob and Rick were broke and exhausted after their six-year struggle—they gave up. It made me sad because they were good men who tried hard. Several interested parties waited in the wings, however. By then *Milagro* had become what some literary pundits were calling "an underground cult classic." The original hardbacks, that I had once so cavalierly given away, were now going for $40 to $60 a pop. My new Curtis Brown film agent, Tim Knowlton (who was eighteen in 1973 when this process began), mentioned that Robert Redford's office had extended a nibble.

One afternoon I found myself in an Albuquerque Sheraton downing beers with a man named Moctesuma Esparza. He was soft-spoken, peered disarmingly out at the world through thick-lensed glasses, and had made only one film that I knew of, a lovely documentary short on a Medenales, New Mexico, weaver named Agueda Martínez; it had been nominated for an Academy Award. Mocte seemed to have little money or clout, which immediately endeared him to me. He had been active in the Chicano movement, which made me like him even better.

The plan was to obtain a NEH grant, in conjunction with the National Council of La Raza, to start the picture. It would be a small-time project geared toward PBS. I would write a script for the Writers Guild minimum. Once we had a script, we'd try to raise major theatrical interest, get a studio involved, and go for a feature film.

Moctesuma and the National Council of La Raza received the bucks to proceed . . . and then my phone rang—it was Robert Redford, interested in the *Milagro* option. Though I explained that Esparza owned it, Bob wanted to visit Taos anyway. By then it was spring 1980, and I was fairly jaded. I was much more involved in cleaning out ditches again, planting the garden, gearing up to face another summer of hysterical fecundity.

In due course Redford called once more. This time he was at a Hudson gas station in Taos, and he was pissed because he'd just been ticketed for speeding while entering town. He had tried to argue his way out of it, but the cop merely smiled and admitted he'd never ticketed a movie star before.

I drove in, met Bob, led him home. We talked at the kitchen table while my kids and various neighbor children peeped in the windows. I reiterated what I'd already explained: Esparza really *did* own the option. Bob laughed: "You mean I drove all the way down here for nothing?" I suggested, "Well, you could always talk with Mocte about it." And he departed after cheerfully posing with my ten-year-old daughter, Tania, and two neighbor friends for a photograph that would decorate their walls for years to come. Months passed. I futzed with a new novel, *The Nirvana Blues* (for which I could not even hustle an advance). Esparza gave no go-ahead to commence a script. He and Bob were talking. Eventually, they decided to work together and signed a joint venture agreement. So now it was Redford with whom I would discuss the script. But Bob had more or less disappeared . . . so finally, as winter approached, I took a deep breath and blammed out a "preliminary" draft screenplay. I considered it a rough document to be used simply for talking points. Hopefully, it would generate reactions so that finally we could proceed.

Mostly, my awkward script generated a big silence.

My initial *Milagro* script was not great shakes—heavy on the sight gags, pretty loose on construction. Bob suggested we now work on it together at Sundance in Utah, during the first annual Film Institute there. He would be in one place for an entire month, and was eager to get started. Me, too.

Then the Writer's Guild went out on strike. I was dumbfounded—*what schmucks!* But I believe in unions and would never cross a picket line. So I would not even chat casually about *Milagro*, let alone touch pencil to paper in search of a better script.

Frantically, during the strike, I churned out a non-fiction book called *The Last Beautiful Days of Autumn*, and scored a modest advance. Redford invited me to Sundance for a week anyway, as a gesture of friendship. I could watch films being made, maybe learn something.

I went, and it was a fascinating and turbulent week during which nobody even mentioned *Milagro*. Shortly after I departed Sundance, the strike was settled. But by then Redford was out there in seventy-two different places at once—to all intents and purposes he had disappeared again. And Mocte was busy cranking up his new film, *The Ballad of Gregorio Cortéz*, plus I was flying around the country on a book tour for *The Nirvana Blues*, and also trying to prepare the *Autumn* book for 1982 publication. Too, I was doing an original script for the Greek director, Costa-Gavras, on the daily lives of nuclear physicists.

So *I* had essentially disappeared to boot!

The next few years were crazy. By the end of December 1982, I'd written several *Milagro* scripts for Esparza and Redford. To help me out, they hired Frank Pierson (the Writers Guild President and author of *Dog Day Afternoon*) to work with me on the *Milagro* project. It turned out Frank and I liked each other and we spent some good times together in Taos hacking away on the script. The Hollywood trades suggested that Redford would soon start filming

*Milagro*. Instead, none of my scripts captured anyone's imagination, and the project, as far as I could tell, puttered off into idleness.

By then I was too busy on other fronts to truly notice. From 1982 until well into 1985, I worked on several different screen projects: another with Costa-Gavras based on a post-nuclear war novel called *Warday*; I also did a rough first draft of *The Magic Journey* for Louis Malle. And, with Karel Reisz, I wrote a screenplay dealing with Haitian refugees, the U.S. Coast Guard, and moral decisions which override unjust laws.

In early 1985, when I was finally dropped from *Warday*, I returned to literature with a great sigh of relief, finished off a non-fiction book called *On the Mesa*, and started a new novel, *American Blood*. At some point around then I learned Redford and Esparza had hired David Ward, Bob's writer on *The Sting*, to do yet another *Milagro* script. Then, suddenly, a draft of Ward's product appeared in my mailbox. When I read it, my heart sank. Its punch line was that a Chicano farmer doesn't know the difference between peas and pinto beans, so he accidentally plants and nurtures a field of peas, thinking they are beans. No such lunacy darkens any page of my novel. And I thought: Oh dear, if *Milagro* is actually filmed according to *this* plot line, now is probably as good a time as ever to commit suicide!

Instead, after living alone for a decade, I married a beautiful woman named Juanita Wolf, and together we went jogging off into the sunset on the mesa west of Taos.

Later, Juanita and I came down to earth and went back to work, together. Part of that work meant attending dozens of meetings called to protest the ongoing "pizzification" of the Taos Valley. As things had been in 1970 when I began writing for the *Review*, so they were still in 1985. Meetings to control rafting on the Río Grande, thereby preserving the delicate ecology of the Wild River Gorge. Meetings to stop a highway bypass that might destroy the last pure agricultural ground around. Meetings to challenge the government's 50-year plan for the Carson National Forest (which compromises nearly 44% of Taos County). Meetings to prevent the Questa moly mine from building a new tailings pond whose effluence could pollute the Río Grande . . .

By the spring of 1986, it seemed Robert Redford was actually going to direct *The Milagro Beanfield War*. *Variety*, the New Mexico Film Commission, and the Albuquerque *Journal* all claimed it was true. The more that rumors flew, the less I believed them. The *Milagro* film project had became a celluloid Amarante Córdova that refused to get made, even as it also refused to die.

A production manager named David Wisnievitz arrived in Taos: we had breakfast at Michael's Café. David was scouting locations. *Milagro* actually . . . almost . . . just about for sure . . . looked like a "go" picture. And the peas at the end of the script had been changed back to beans. *That* was a relief! But I still figured the odds were a million to one against it. And I was broke again, so I took another film job, a six-hour TV miniseries on the life of Pancho Villa and the Mexican Revolution.

When Redford next called, he informed me that Universal planned to front the ten-million epic, and would I do a fast polish of the script? Of course I would. Juanita and I traveled to Sundance, meeting with Bob and an assistant, Sarah Black, for three days in June. Back in Taos, over two horrible, manic weeks, I churned out a polish of David Ward's script.

Exhausted, I completed the script, and a courier arrived to grab my work and pouch it off immediately to Bob. I expected an instantaneous reaction, then more day-night frantic working sessions to bang it into shape before filming started. Instead, the script disappeared out there, and I heard next to nothing for three weeks: no phone calls, utter silence, zilch. I couldn't worry about it too much, however, because by then I was frantically trying to produce a "bible" for the Pancho Villa films.

Soon the starting day for the *Milagro* shoot loomed just two weeks away. All the New Mexico papers flaunted the *Milagro* follies. Just north of Santa Fe lies a small town, Chimayó, where the main sets were to be built. But at the last minute, Chimayó rejected the movie. Residents of the old plaza claimed the movie would shatter their privacy. And for a moment it looked as though the project might collapse. Instead, the entire operation relocated to another, more remote, village called Truchas.

Of course, Truchas is perched at 8,500 feet, where the first snow is liable to fall on August 15[th]! Oblivious to such facts, the set makers started all over, with orders to build Rome in a day. In late July I asked David Wisnievitz how it was all going. He replied "No problem." Considering that the shoot must begin in two weeks, that they had not even finished the Truchas sets, that there was no cast as yet, that the script was a total question mark, and nobody really knew where Redford *was* . . . things couldn't have been more hunky-dory.

Incredibly, Bob finally appeared, rented a house in Santa Fe, and began filming *The Milagro Beanfield War* two months later than it should have started, and still without a cast. I did a hasty polish of a script compiled by Sarah Black (and a fellow named Jim Parks) out of former versions by David

Ward and myself. I understood that another writer on the set was doing an all-new, computerized version. Frankly, I was a trifle befuddled by all the chaos surrounding the launching of such a mammoth enterprise. But by then I was so busy editing my new novel, and slaving away on Pancho Villa, that I couldn't worry much about *Milagro*.

Though Bob moved slowly at first, the dailies Juanita and I saw were full of life and humor. The actors and crew were very friendly the few times we managed to visit the set. The eleventh-hour cast was varied, rich, funky, cheerful: Chick Vennera, Ruben Blades, Christopher Walken, Trinidad Silva, Daniel Stern, John Heard, Sonia Braga, Julie Carmen, Freddy Fender, James Gammon, Melanie Griffith, Carlos Riquelme, Robert Carricart . . . ad infinitum. Fifty-three speaking parts! My favorite was a local Santa Fe woman whom Redford apparently just spotted on the street. Her name is Julia García and she plays a daffy old lady, Mercedes Rael, who flings pebbles at everybody. She seemed absolutely perfect.

By October I had grown quite complacent about the whole process. Then— *boom!* On October 22, the producers of a film about the life of land grant activist Reies Tijerina, to be called *King Tiger*, filed suit against the *Milagro* film, against Redford, and against script writer David Ward, claiming that *Milagro* was based on Tijerina's life. They wanted an injunction against the continued filming of *Milagro*. They claimed our film had caused Columbia to shelve *King Tiger*.

Naturally, I was flabbergasted. José Mondragón, a 35-year-old unemployed handyman who cuts water illegally into a half-acre beanfield bore some kind of resemblance to an evangelical preacher who became the messianic leader of a land grant movement that climaxed in an armed raid in the Tierra Amarilla, New Mexico courthouse in 1967? Wow! It had to be a joke. A bad joke. Though my book had often been vilified in the fourteen years since its publication, no one had ever suggested it was based even remotely on Tijerina.

Neither I nor the book (at the date of this writing) were being sued. But I was rendered nearly catatonic by visions of negative publicity: for Tijerina, for myself, for *Milagro*, for the whole effort to create a project with dignity, with political and cultural integrity. It was an insult to all my friends, and to the local struggles which had inspired the novel.

That night, reporters phoned for my reaction. I could think of nothing to comment except that the whole thing was absurd. Privately, I believed the suit was a tragedy. I figured the upcoming scandal would undermine a trust I felt I'd built up over seventeen years of organizing in, and writing about, northern New Mexico. I wished to tell reporters: "To say *Milagro* is based on Reies Tijerina is like saying Walt Disney's *Pinocchio* is based on the life of Christ." Fortunately, I managed to keep my mouth shut.

Still, I envisioned a thousand obscene repercussions against which I'd be helpless to act. No doubt, the suit would be the most offensive intrusion on my life—and on my work—that I had ever experienced.

But nothing happened. Gory details were announced to the press, but the legal papers, so far, anyway, have not been served. After one day the newspapers dropped the story. They picked it up briefly, a month later, when Tijerina pulled out of the suit. Then all fell silent again. And *Milagro* continued filming as if the suit didn't exist. Once again, Amarante Córdova had refused to die!

Came finally the last day of the shoot. Good friend Victoria Plata, who handled extras on the film, asked Juanita to be in the scene. We stumbled out of bed at 5 A.M. and slouched down the highway to Santa Fe in a pickup that skidded dangerously on sheer ice—it had just snowed eighteen inches! Bad weather had held up production by more than a month.

We spent a long day in the posh living room of a ritzy house just north of Santa Fe. Mostly, we waited around. Dozens of huge semi trucks got stuck on the narrow, muddy roads leading to the house. An army of technicians tromped about in muddy moon boots and puffy ski jackets, laying down a million miles of wires, setting up klieg lights and gelatin sheets, and in general destroying the lovely house and its pretty grounds.

The scene was supposed to take place in summertime. Half the extras had frostbite on their toes. The shoot lasted from 10 A.M. to 9 P.M. Juanita stood around looking radiant, sipping apple juice "champagne," loving every minute. I reacted more wearily: How did such a ponderous operation ever accomplish *anything? Eighty thousand dollars a day?* Imagine if I could send that loot in medical supplies to the Sandinistas in Nicaragua! Many times the mariachi band of Debbie Martínez—La Chicanita—struck up its peppy music for the cameras, and that helped keep people awake. Redford chewed gum and stayed in remarkably good humor through it all. Some crew members were real glad it was almost over, however: recently, graffiti had sprung up around Santa Fe—"Free the Milagro 100!"

By day's end, some extras grumbled that it was a lot harder than they had ever thought it could be. But not Juanita. She looked more ebullient at 9 P.M. than she had at ten that morning. Hey, she actually had been directed by Robert Redford! And from now on her bags would be packed, just waiting for Hollywood to call.

Yes, Virginia, there really *is* some glamour in making movies after all! What, then, remains as the moral of this story?

In 1957, a boy wanders among the Chiricahua Mountains . . . and nearly thirty years later a famous actor directs a movie whose roots travel back to that bewitching summer. Did I write *Milagro* and the other books excerpted here for the same reasons that Esparza and Redford risked more than ten million dollars to launch the film? Probably not. *Milagro*, and my other work, is inspired by a radical political dream which is not often touted in our culture. Yet, hopefully, the film will reflect honorably upon the land and the people who work it, not just in Taos or New Mexico, but everywhere. I figure if the film can hold even half a candle to *Salt of the Earth*, to Andrés Martínez, or to the values of many of my Taos neighbors, it will make me very happy indeed.

I have been grateful for Redford's pronouncements to the media. On November 2, 1986, the Philadelphia *Inquirer* quoted Bob as follows:

> *First,* Milagro *is about the little guys against the big guys. Second,* Milagro *also addresses the issue of responsible land use, which is an interest of mine. A point that developers don't always see is that their new projects often threaten to eradicate indigenous culture. If this goes on unchecked, places that have distinct ethnic and architectural heritages are in danger of being homogenized, being like every place else. It's not widely known that the Hispanic community in the valley has a history predating Plymouth Rock, and the Indian community three times that. And that's the third reason it's exciting for me to work on* Milagro. *Like* Jeremiah Johnson, *it teaches history in an entertaining way. I think there's a suspicion among many Americans that the peoples of northern New Mexico are not real citizens. The truth is, they were here long before us Anglos. My criticism of this country—particularly under the Reagan administration—is that we're ignorant of history and other culture, and I think* Milagro *might help correct that cultural ignorance.*

I trust that in the *Milagro* film will shine all the strengths of this lovely territory, all the humanity of its denizens, all the concerns they have for the future of the Taos Valley. Most importantly, I hope that the movie, and this book will help add to a wider awareness of the need to protect what remains, and to change what isn't working coherently for us now.

For what remains, always, is a fragile beauty that we can ill afford to squander.

## 11

# For the Sake of Peace in the Valley: The Negotiated Settlement in the Taos Water Rights Adjudication

### Sylvia Rodríguez

On May 30, 2006 a public ceremony was held at Taos Pueblo to celebrate the official signing of a negotiated settlement in the water rights adjudication of the Rio Pueblo de Taos, known as the Abeyta case. The two-hour event took place outdoors in the north plaza of the pueblo, near the banks of the Rio Pueblo, framed against the spectacular backdrop of Taos Mountain. Tribal elders dressed in beaded finery sat alongside elected officials and representatives of the other parties to the suit, beneath a ramada shaded with freshly cut greens. The pueblo governor's canes of authority lay on a long, brightly blanketed table where the signing would be done. White awnings erected on either side shielded audience members from the blazing sun. Next to the podium, U.S., New Mexico, and Taos tribal flags flapped wildly in the wind. A succession of speakers preceded the actual signing, starting with the pueblo *cacique* or spiritual leader, the pueblo governor, the mayor of Taos, the governor and the attorney general of New Mexico, and representatives from each of the negotiating parties: Taos Pueblo, the State of New Mexico, the Taos Valley Acequia Association (TVAA), the Town of Taos, El Prado Water and Sanitation District, and twelve Mutual Domestic Water Consumers Associations. The audience consisted mostly of family members of the men and attorneys who had participated in the long and arduous negotiation process. At least half a dozen members of the press crouched in a wide arc facing the podium against the hot, unrelenting wind.

After thirteen pages of signatures were completed, everyone joined in a round "friendship dance" circling a chorus of drummers. The crowd then walked to the community center across the river, where it was treated to a generous feast prepared by pueblo women. Dessert featured two large decorated cakes commemorating the historic event. The signing ceremony concluded a ten-week period that began on March 17, when after seventeen years of extremely difficult and complex negotiations, the parties finally reached an agreement. The settlement was announced at the end of March and posted on the Office of the State Engineer's (OSE) website. In April the OSE held a press conference and two public meetings in the town of Taos to summarize the terms of the hundred-page document. Meetings to inform the tribal council and the community at large were also held at the pueblo.

The flurry of publicity during the spring 2006 disclosure period followed almost two decades of dead silence surrounding the negotiation process as well as the case itself. The purpose of the Abeyta case, filed in 1969, is to adjudicate all the water right claims of Taos Pueblo and non-Indian claimants to the Rio Pueblo, Rio Hondo, and their tributaries as well as groundwater in the Taos Valley. At stake is virtually everyone's legal right to use water: how much, for what purpose, and in what order of priority. Even though water is a renewable resource, every drop of it in Taos as well as elsewhere in the state is already appropriated. This means that there is no extra water available to accommodate new demands, and an increase in one user's supply requires a corresponding reduction in someone else's. It is a zero-sum game.

Or such was the case until the 1960s, when "new" water was imported from the Colorado River Basin across the Continental Divide into the Rio Grande and its tributaries. This importation of water from a legendary western river that does not even flow through New Mexico is known as the San Juan–Chama Diversion Project, because it channels Colorado River water into the state via the San Juan and Chama tributaries. It has involved many millions of dollars' worth of mostly federally funded hydraulic construction projects including an enormous pipeline drilled through a mountain and the expansion of a system of reservoirs to store and release imported water. It brought new water into the system that could be acquired by government contract, and thus enable all future growth along the Rio Grande corridor. It was a game-changer that triggered New Mexico's water adjudication suits, including Abeyta and the better-known, intensely controversial Aamodt case. This is because the principle of prior appropriation implemented in the American West decrees "first in right, first in time," meaning that those with the oldest water rights have priority over all junior users, and during times of scarcity they get first take. Adjudication requires that every water right claim be determined as to ownership, location, use, quantity, and priority ranking. Priority and demand determine market value. Adjudication is thus a crucial step in the complete rationalization of the system, in order to "measure, meter, and market" every drop of the most precious, limited, and limiting resource in the desert borderlands of North America.

The adjudication lawsuits oppose each of the Rio Grande Pueblos' aboriginal water right claims of priority to all junior claims, starting with the acequias or irrigation communities established during the Spanish colonial era. Each case is different, enormously complex, politically volatile, and glacially slow. Aamodt, opposing the pueblos of Pojoaque, Tesuque, Nambe, and San Ildefonso against all their non-Indian neighbors, is said to be the oldest ongoing case in the U.S. court system. Abeyta is only two years younger. The adjudications started to enter an active phase in the 1980s as population growth and urban development in the region intensified demands on water. As they prepare for and enter litigation and face myriad simultaneous challenges to the maintenance of their land base, water rights and irrigation traditions, acequia communities have grown self-conscious and eloquent about their place-based ethnocultural identity. Like the pueblos, acequia proponents deploy increasingly sophisticated methods to defend their resources and traditions. Acequia associations have formed coalitions within and across watersheds as well as statewide in order to share information, fundraise, support and educate members, and lobby for collective interests. The TVAA, representing fifty-five Taos area ditches, is such an organization.

The case of Taos is typical in some respects and unique in others. As in all adjudications, claims to and disputes over local water between Taos Pueblo and its neighbors go back roughly four hundred years. Existing documents provide a vivid but foreshortened and partial record mined by scholars largely for purposes of litigation. In order to grasp the complex particularity of Taos or any other case, we must first look at the lay of the land, including where and how the waters flow. Think of the Taos basin as a handful of rivers flowing southwest into the wrist of the Rio Pueblo. Using your right hand and moving south to north, let the thumb be the Rio Grande del Rancho (RGR), made up of two streams (Rio Chiquito & RGR) coming together; your index finger is the Rio Fernando; your middle finger is the Rio Pueblo, fourth the Rio Lucero, and little finger the Arroyo Seco. In reality there are six fingers in the larger geographic "hand" of the Taos basin, including the Rio Hondo several miles to the north, a tributary of the Rio Grande but not the Rio Pueblo. The Rio Hondo adjudication (Arellano) was incorporated into Abeyta because its waters mingle with the Arroyo Seco and Rio Lucero via a certain unusual ditch and the pueblo claims aboriginal use.

The first permanent agricultural settlement we know of in the Taos basin was Taos Pueblo on the upper banks of the Rio Pueblo. All those that developed after European and prior to American conquests clustered along the rivers in order to divert water into fields. The precise sequence of colonial settlement in the Taos Valley is not archaeologically known, but the thumb, forefinger, middle, and ring fingers all had acequias irrigating off them prior to the 1680 Pueblo Revolt. Irrigation from the Arroyo Seco, Rio Hondo, and San Cristobal rivers came when a burgeoning population pushed northward in the late 18th and early 19th centuries. Over the centuries, Taos Pueblo managed to fend off *vecino* encroachment only from the upper banks of the Rio Pueblo, around the village and its contiguous farmlands, and upstream all the way to the source of the watershed, Blue Lake and the surrounding mountain wilderness. Because of its upstream location, the pueblo has always enjoyed undisputed first take from the Rio Pueblo. By 1730 the pueblo was protesting encroachment by settlers on lands contiguous with the village, and by 1800 was attempting to buy back tracts along the upper banks of the Rio Lucero. Legal dispute over the waters of the Rio Lucero erupted around 1815 when Arroyo Seco was settled and water began to be diverted upstream from both pueblo irrigators and vecinos in Los Estiercoles or El Prado. How the water of the Rio Lucero has been or should be divided among these three communities constitutes a core issue between the acequias and the pueblo in the negotiated settlement. Customary division of the Rio Pueblo between the pueblo and downstream communities, including the town of Taos, is another. Also important to the pueblo is the large, fertile wetland west of the Indian village known as the Buffalo Pasture, located near El Prado between the Rio Pueblo and the Rio Lucero.

Each of the parties to the settlement has its own particular set of concerns, some of which relate to their capacity to grow and where the water will come from. The town, the pueblo, and the El Prado Water & Sanitation District aim not only to secure their existing claims but also to acquire more water for growth. The Mutual Domestic Water Associations seek additional water to meet existing needs and remedy inequities in their current supplies. Only the acequias struggle simply to keep what they have. Acequias have the second oldest priority rights, but unlike pueblo rights they can be alienated or purchased, and are therefore *the* prime target for water speculators. The complicated terms of the settlement can be characterized but briefly. One negotiator called it a "win-win" solution from the acequia standpoint because of three attributes: not one *parciante's* existing water rights would be sacrificed; water-sharing traditions (with the pueblo) on the Rio Pueblo and the Rio Lucero would continue; in exchange for acknowledgement that *all* of its rights are aboriginal, the pueblo promises *never* to call for priority administration on either the Rio Pueblo or the Rio Lucero.

The negotiation seeks an alternative to costly and potentially endless litigation on a thousand big and minute issues encompassing virtually every water dispute that has persisted for generations between Indian and non-Indian users. The pueblo claims historical rights to several thousand more acre-feet of water than it currently uses, and agrees to "forebear" in their exercise until a corresponding amount can be acquired by them through purchase so they can be retired, in order to offset the deficit caused by increased pumping. Key to the settlement is the construction of numerous wells in the Taos basin for storage, pumping groundwater, and "mitigation: to offset the future surface water depletion effects on Taos Valley tributary streams that result from groundwater production."

In the end, what made it possible for negotiators to reach an agreement on each and every issue is the prospect of drilling new wells to provide a few thousand acre-feet of San Juan-Chama water. But because Taos sits north of the point where actual imported wet water enters the Rio Grande, this "new" water has to come out of the ground. The depletion of groundwater inescapably depletes surface water, and vice versa. One negotiator explained the underlying logic this way: "the Taos Basin loses much of its water to the Rio Grande each year. The increase in future uses are made possible because of this net excess water available to the Taos Basin. San Juan-Chama water is needed to offset the increased depletions on the Rio Grande caused by increased future groundwater pumping." Thus a profound irony inheres in this grand hydraulic shell game: San Juan-Chama water is being used to solve the very dilemma triggered by its introduction into the Rio Grande system. It boggles the mind.

This in no way diminishes the impact Abeyta and other adjudications have had on the day-to-day lives of real people. Lifetimes have been consumed and shaped by water right—and land grant—cases in northern New Mexico. One is reminded of Dickens' novel, *Bleak House*. Consider that the negotiation process took seventeen years, in which at most a handful of individuals participated from start to finish. These included one or two men from Taos Pueblo and two from the TVAA, one of them the attorney. Other representatives of all the parties came and went: changed jobs, retired, left office, or died. Soon after initiating discussions in 1989, the TVAA and pueblo agreed to maintain strict confidentiality and keep the story out of the newspapers because publicity would endanger the process. Over the years the negotiations would stall, spurt ahead, backtrack, rupture, dwindle, and grind on. Throughout this time everyone was also preparing for trial. In 2003 a professional mediator was brought in, which accelerated the process. During the final three years, the TVAA, pueblo, and other party representatives met every other week, sometimes for several days, all day. These included bilateral and multilateral meetings. They were held both in Taos and Santa Fe.

Testimonies by individuals involved in these struggles give a human face to what otherwise seems a remote, esoteric, abstract process. When the agreement was finally reached in 2006 and the OSE held a press conference, the atmosphere of relief, elation, and exhaustion was palpable. It was on this and ensuing occasions that participants spoke candidly about the import and magnitude of what they had managed to accomplish. Some described how stressful and contentious the process had been, alluding to occasions when people walked out or said things they later would regret. Pueblo and acequia irrigators learned facts about each other's day-to-day lives and ways of doing things that surprised and moved them. Negotiators explain how, over the years, they grew to appreciate individuals on the other side and, in the end, realized how much they had in common. One pueblo man remarked that he came to trust his parciante counterparts because, like him, they wore jeans, not suits, and after their endless meetings would go home to feed their livestock or clean out a ditch. The mix of intimacy and distance is striking between these neighbors who have lived for centuries along the same rivers, sharing and fighting over the water.

Once the agreement was reached, everyone, including the State of New Mexico, had a stake in moving it forward to the next phase: asking Washington to adopt legislation and provide federal funding for the settlement. In the words of a flier issued jointly by Taos Pueblo and the OSE, "The unity and agreement demonstrated by this proposed settlement is unprecedented and the parties will provide this message to the State's lawmakers in Washington in the next phase . . ." All players emphasize that this unique and hard-earned settlement will allow the local parties to co-exist peacefully in the Taos Valley.

Who the winners and losers are will depend on what happens next. The Abeyta settlement bill passed the House in 2009 and, as of this writing in 2010, is still pending before the Senate amidst efforts to package it with other Indian water right settlements including Aamodt, in an omnibus bill geared to draw broad bipartisan support. President Obama signed a letter in support of the settlement in September 2010, but the Senate adjourned early so members could campaign for a mid-term election. The original federal funding request for Abeyta has been cut from $133 million to $121 million, the lion's share going to the pueblo for water development ($58 million) and infrastructure ($30 million). Most of the remaining $30 million would go to the Town of Taos, the Mutual Domestics, and the El Prado Water and Sanitation District (EPWSD) for water development and mutually beneficial infrastructure including the mitigation wells. The State of New Mexico is supposed to contribute about $20 million. These figures and the situation in general remain fluid and subject to change. The original settlement specifies an expiration date of December 31, 2015 for congressional approval and signing by the President. If conditions are not met, the agreement shall be null and void, and any unexpended federal or state governmental funds, together with any income earned thereon, and title to any property acquired or constructed with expended federal and state funds, shall be returned to the appropriating entity, unless otherwise agreed to by the Parties in writing and approved by Congress or the New Mexico Legislature, whichever is appropriate depending on the source of funding, with the exception of the $250,000 that was appropriated by the 2005 New Mexico Legislature for water rights for the EPWSD, which shall not have to be returned to the state.

In short, the whole thing could unravel, leading to ever more costly litigation. Then, as the saying goes, only the lawyers are sure to win. A final irony is that the least expansionist party, the acequias, is most at risk. In an era when multiple forces conspire to undermine their ability and inclination to do so, parciantes must continue to irrigate or risk losing their rights through forfeiture or transfer. The pueblo's stated intention is to acquire and ultimately retire these rights. At the same time, non-Indian water brokers and metropolitan users covet them for purposes of profit and ever expanding need. Because of its protected federal trust status, pueblo water cannot be sold or forfeited through non-use, although it can be leased to non-Indian users, which may well become part of a future economic development strategy. Irony thus portends a larger tragedy that derives from a socially pervasive, ultimately unsustainable, economic policy whereby those who live beyond their means are rewarded, at the expense of those who have learned to subsist within them.

## Epilogue

On Wednesday, December 8, 2010, President Obama signed into law legislation that settled a number of Indian water rights cases, including both the Abeyta and Aamodt adjudications. The bill had finally been passed by the House on November 30. The measure included $66 million for the purchase of water rights and construction and water management projects in the Taos Valley. It authorized an additional $58 million in future funds needed to fully implement the terms outlined in the settlement. The state of New Mexico is expected to contribute approximately $20 million toward the effort. Upon passage and signing of the bill, Taos Pueblo, the TVAA, and other parties to the suit heaved a collective sigh of relief. The case had lasted for forty-one years. March 31, 2017 has become the new deadline for full implementation of the settlement. Ever the realist, the president of the TVAA was quoted in the *Taos News* as saying the future work of implementation is now cut out for them.

# Water Transfers and the Weight of Public Welfare Considerations

## Calvin Chavez

Foreword and Afterword by John W. Hernandez and Linda G. Harris

**Foreword**

*C*alvin Chavez's concise account of the Ensenada Ditch Association's 1982 legal battle to preserve its community water rights contrasts to its more dramatic staging in the headlines, in John Nichols' novel *The Milagro Beanfield War*, and in Robert Redford's movie of the same title. Why has the issue of community water rights proved so fascinating? The answer lies somewhere between New Mexico's land grant past and its evolving water laws.

The story of the Ensenada Ditch Association began in 1832 with the Tierra Amarilla Grant to Manuel Martinez, his sons, and a few other Chama River families in northern New Mexico. (The communities of Los Ojos, La Puente, Los Brazos, Ensenada, and Tierra Amarilla today occupy lands in the former grant. Tierra Amarilla later became the Rio Arriba County seat.) Under the 1856 Treaty of Guadalupe Hidalgo, Francisco Martinez, one of Manuel's sons, received federal title to the land as a private grant. He later sold the granted lands to Thomas B. Catron, a major land baron and political power. In 1883 Catron received formal ownership through a quiet title, which gave him ownership of the grant except for the "informal conveyance of some very small pieces of land." These exceptions were known as the "Catron Exclusion," and today are the irrigated lands served by the Ensenada Ditch.

The Ensenada Community Ditch takes its water from the Brazos River, a major Northern New Mexico tributary of the Chama River and from Nutritas

Creek. The Ensenada is a relatively long ditch by acequia standards, weaving in and out of the south line of the Catron Exclusion. The lands served by the ditch include 982.1 acres with a claimed use of one and a third acre-feet of water per acre. This diversion totals about 1,300 acre-feet of water per year. The Ensenada acequia delivers water to a few large tracts of land, as well as to many long, narrow farms that head on the ditch. These narrow holdings are characteristic of northern New Mexico acequia lands. Many once were larger farms, but over time they have been divided into smaller parcels as inherited shares, with each farm entitled access to the acequia.

While Ensenada ditch commissioners reported in 1936 that "crops [were] generally good and [there was] sufficient water supply," other community ditches were being absorbed into larger entities. In the 1920s, the acequias on the Rio Grande near Las Cruces became part of the Elephant Butte Irrigation District, while a decade later, ditches in the Middle Rio Grande met a similar outcome. By 1950, the acequias in Santa Fe had ceased to exist altogether. Still, none of these moves brought the protest and notoriety that would come with the challenge to the Ensenada Ditch owners.

No doubt with these lost acequias in mind, the Ensenada community ditch owners in 1982 filed a protest against the transfer of water rights from their community ditch for use in a commercial development. The water rights and land owners, by then numbering 1,000, lived on narrow strips of farmland, and considered their acequias in every sense, "community ditches." The local farmers who maintained and cleaned the ditches believed if a land and water right owner on an acequia sold his lands and water for a purpose other than farming, the remaining farmers would also feel the loss of the farmland. The shared duties of ditch cleaning and maintenance would weigh heavier on the reduced number of people to do the work. Just as important would be the effects of losing the traditional "commons" in the original land grant. Activists protesting the sale of the Tierra Amarilla Land Grant triggered the June 1967 raid on the Tierra Amarilla Courthouse where they demanded the return of their ancestral land grant. Fifteen years later, the Ensenada Community Ditch members protested the loss of their land and water by waging their own "water war."

**Developers Versus the Ensenada Ditch Association**

In New Mexico, the law requires that the state engineer must take the public interest into account when considering applications for water rights transfers. The law applies in instances where the proposed appropriation is for beneficial use, when the application is for transferring water rights, and if the application changes any component of the right. This explicit authority was passed by the New Mexico Legislature and enacted into law in 1985. The court case that spurred the legislature to act was a water rights application filed in 1982 by Howard Sleeper and Hayden and Elaine Gaylor to transfer 75.5 acre-feet of their water rights on the Northern New Mexico Ensenada Ditch to Tierra Grande, an area recreational development.

Ensenada is a small picturesque farming community located two miles north of Tierra Amarilla, the Rio Arriba County seat. A raid on the courthouse in 1967 drew national attention when armed protestors led by Reies Lopez Tijerina took over the courthouse in an attempt to restore New Mexican land grants to the descendants of their Spanish colonial and Mexican owners.

Tierra Grande, Inc. and Peñasco Ski Corporation contracted with Sleeper and the Gaylors to purchase water rights. The developers planned to use the water for their recreational developments and for an associated subdivision on the Tierra Amarilla Land Grant near Ensenada. To obtain materials for subdivision roads, Tierra Grande had removed gravel from the bed of Nutritas Creek. The developer also had built a dam across the Nutritas to form a small lake. When the state engineer learned that the new reservoir held more than 10 acre-feet of water, an amount that required a state permit, he ordered the dam breached, making it unusable. The developers complied with his order. Then on September 17, 1982, Howard Sleeper and Hayden and Elaine Gaylor filed a water rights application with New Mexico State Engineer Steve Reynolds, proposing to transfer some of their surface water rights, which historically had been used to supply irrigation from the Ensenada Ditch to Tierra Grande Inc. The developers needed the rights to fill and maintain water in their newly formed recreational reservoir.

Early in the irrigation season, Nutritas Creek runs full with snow-melt, water which is diverted to the Ensenada Ditch. Later in the season, when the spring runoff in Nutritas Creek is insufficient to meet the irrigation demands under the Ensenada Ditch, water is diverted from the Rio Brazos. The flows of the ditch are used for irrigating small gardens and to pastures that the community uses to graze their livestock. The application was protested by members of the Ensenada Ditch Association, the nearby Los Ojos Community Ditch Association, and the domestic water associations of Ensenada and Los Ojos. Under New Mexico law, any proposal to change an element such as a point of diversion, place, and or purpose of use of an existing water right must first be published in the legal section of a newspaper in the area. The purpose of the notice is to inform the public of the proposal and gives anyone the opportunity to protest the proposal if they believe that approval would be to the detriment of existing water rights. After the community associations filed their protests, the applicants modified their application to reflect their intent to change the place and purpose of use of just 61 acre-feet of water for one-year to fill the lake, then add 13.3 acre-feet each year thereafter to maintain the level of the lake to account for evaporation. The original application specified a permanent transfer of 75.5 acre-feet to the development.

The protestants expected that under these changes, the state engineer's appointed hearing examiner would accept the transfer, therefore they presented no evidence to support their claim that the application should be denied. The evidence at hearing in fact supported the conclusion that granting the application would not cause impairment or be detrimental to existing water rights. The state engineer adopted the hearing examiners recommendation and on January 27, 1984 approved the water rights transfer.

In response, the protestants, who had anticipated the state engineer's ruling, appealed his decision to the district court. The matter was assigned to First Judicial District Court Judge Art Encinias for hearing on March 28, 1985. The hearing was held in the same Tierra Amarilla Courthouse made famous by the armed raid by land grant dissidents in 1967. The hearing was packed. On July 2, 1985, Judge Encinias issued his ruling reversing the decision of the state engineer and denied the transfer. In his ruling, Judge Encinias found that the state engineer erred in his finding that the granting of the application would not be to the detriment of existing water rights, but rather that in early spring livestock would be deprived water derived principally from Nutritas Creek and the Rio Brazos. Encinias also found that the water users would be deprived their first irrigation watering in the spring. The spring watering benefits the land by moistening the soil in preparation for planting and the water fertilizes the soil by providing rich silt carried by the waters of Nutritas Creek. The settlers in the area named the creek Nutritas because of the rich mineral nutrition carried by the waters onto the irrigated fields. The judge went on to say that the state engineer must consider the public interest in transfer or change decisions, whether or not articulated in statute for the reason that the state engineer is a public official charged with the supervision of an important resource belonging to the public. The state engineer may disapprove the application for transfer or change if it is contrary to the public interest. The Sleeper application was contrary to the public interest, the judge ruled, because the Northern New Mexico region possesses significant history, tradition, and culture of recognized value not measureable in dollars and cents in which the relationship between the people and their land and water is central to the maintenance of that culture and tradition. The imposition of a resort-oriented economy in the Ensenada area would erode and likely destroy a historically distinct culture.

At the hearing, Judge Encinias was the first to point out that the concept of protecting a resource from the influence of monetary gain never had been factored in when considering water rights matters in the past. Instead, the only serious consideration had been the influence of water depletions in consideration of impairment thresholds. The local community believed that Encinias' public interest decision in denying the transfer of water from the ditch to a new non-irrigation use was correct.

However, the district's judge's decision was appealed to the New Mexico Court of Appeals and in 1988 the court reversed the decision ruling that Encinias erred when he denied the requested transfer on the basis that it was contrary to the public interest. State law in force at the time the application was filed did not allow denial of the proposed transfer on the grounds of public interest considerations. The New Mexico legislature amended the state law in 1985 to include public interest standards to apply to water rights transfer applications. The court further found that contrary to claims by association members, "water rights do not include a right to receive a traditional or historical amount of silt carried in the water." The transfer of water from the Ensenada Ditch system to fill Tierra Grande's recreation lake was approved. In the decision, the appeals court ordered the district court to include a provision for monthly metering of the water diverted to the dam in the state engineer order.

**Afterword**

In the years since the Encinias decision, there have been a number of legislative efforts to prevent or at least limit the transfers of acequia waters to uses other than irrigation under a community ditch. The state legislature has always treated the administration of acequia water rights somewhat differently than those involving other water right issues. The legislature requires that each ditch have an elected set of commissioners and an appointed "mayordomo" who is the chief executive officer for the acequia association. There are many sections in the statues assigning the "mayordomo" significant powers and listing his duties.

Recent statutory changes have added to the powers of the commissioners and the "mayordomo" and made it extremely difficult to transfer water from a community ditch to any use other than irrigation. In 2003, the legislature adopted a set of conditions that make water right transfers out of community ditches virtually impossible. These requirements included the following:

> that the commissioners for acequias adopt procedures, approved by the ditch member, dealing with the transfer of water rights for uses other than irrigation;
> that any application for a change in the place or purpose of use of acequia waters must be referred to the ditch commissioners for approval prior to consideration of any change by the state engineer; and
> that the commissioners of a ditch may deny the application for change, if they find that the change would be detrimental to the acequia or its members, and if the commissioners render a written decision explaining their rational.

An applicant, seeking to make a change in the purpose of use of water from a community ditch whose application is denied by ditch commissioners, does have the right of appeal, not to the state engineer, but to district court. An aggrieved member of the ditch association also enjoys this route of appeal. The district court can only reverse the ditch commissioners ruling by finding that they acted fraudulently, arbitrarily, capriciously, or not in keeping with the law. It is not likely that, in the future, that there will be many transfers of water rights out of community ditches in New Mexico.

# 13

# Conflicts in the Division of New Mexico's Share of the Colorado River

John W. Hernandez

### An Outsider's View of the Colorado Water Allocation Process

The quote below is an outsider's somewhat fictitious view of the process that the state engineer used in allocating New Mexico's share of the Colorado River water to assorted development plans. It is taken from John Nichols' 1974 novel, *The Milagro Beanfield War* that chronicled local opposition to part of the original scheme for using some of the Colorado water in the Taos area for supplemental irrigation and, in this case, for newly irrigated lands. Nichols gave then New Mexico State Engineer Steve Reynolds, and his lawyer, Paul Bloom, fabricated names, but many of the characteristics assigned to the two, fit their real-life persona.

> "[They] knew more water law than the rest of the state put together.... [He] was more responsible than any other person or group for what water the state had obtained... through interstate compacts and reclamation projects. During that time they had weathered the heaviest political storms to sweep the state. They had also sweated, plotted, finagled, begged, twisted, and driven their way to what they felt was the state's fair share of Colorado River basin water; they had made deals with Texas and California, with Arizona and Colorado and Utah and they had created lobbies in Washington to have dams built and rivers channeled; they had set into motion adjudication suits to determine how much water

people did or did not have in all areas; they had literally decided [where] the rivers would run and which people must benefit the most from those rivers."

## Pre-War Basis and Origins of the San Juan-Chama Project

The plan for a trans-basin water transfer from the San Juan River into the Rio Grande may have had its genesis in the development of the 1922 Colorado River Compact between the Lower Basin states of Arizona, Nevada, and California and the Upper Basin states of Wyoming, Colorado, Utah, and New Mexico. The 1922 compact appeared to promise New Mexico a relatively large amount of water for future use, perhaps the entire flow of the San Juan River.

At the time of the compact, considering the average volume of its flow, there was almost no use of the San Juan River in New Mexico. The San Juan River begins in the mountains of Colorado and flows southwesterly 50 miles or so and then westward to the northwestern Four Corners area where the states of New Mexico, Colorado, Utah, and Arizona meet. The towns in route were Aztec, Bloomfield, Farmington, and Shiprock, all small in 1922. That corner of the country was home to three Native American tribes: the Navajo, the Southern Ute, and the Jicarilla Apache. By any measure, the region was under-populated when the compact was drawn.

The 1922 compact left decisions on the division of the water awarded to the Upper Basin states for another day. Arizona refused, until 1944, to ratify the Colorado River Compact and that, and along with the impacts of the Great Depression, put a hold on efforts to develop New Mexico's still unknown share of Colorado water. At the time of the compact, Colorado was also pushing for a compact on the Rio Grande that would involve Texas, New Mexico and Colorado. In 1929, an interim Rio Grande Compact was reached and, interestingly, it contained language providing for the possibility of trans-basin diversions of Colorado River water.

Progress in the utilization of our Colorado River water was slow despite the existence of an idea that would help solve shortages on the Rio Grande in New Mexico, where the supply of the river had long been fully appropriated for various uses. Growth and economic development were taking place in the central corridor of New Mexico and that was where the state needed to use a significant share of the promised Colorado River water. That water would have to come from the San Juan River, New Mexico's sole-source of Colorado water. Thus, a project to bring Colorado water across the Continental Divide was bound to be supported.

Pre-World War II planning did not focus on water for the Navajo Indian Irrigation Project or for the Jicarilla Apache Reservation, although both groups held large tracts of land on the western side of the Continental Divide. It is a little surprising that no strong support had developed at that time for use of a significant part of New Mexico's anticipated Colorado water in the economically underdeveloped northwestern edge of the state. Maybe the western-slope residents just assumed that a major share of San Juan water would be theirs some day. Early on the Bureau of Reclamation initiated studies of the possibility of diverting San Juan River waters into the Rio Chama that enters the Rio Grande near Hernandez, New Mexico. In 1933, preliminary field work began in what would become the 1936 Rio Grande Joint Investigations. The first official action in New Mexico was in 1934 when Governor Andrew W. Hockenhull created a temporary planning board to consider, among other natural resources issues, the transport of water from the Colorado River system into the Rio Grande. This unofficial board proposed, among other things, a San Juan-Rio Grande diversion project. In 1938, a final version of a Rio Grande Compact was arrived at, and again there was specific language with respect to trans-basin diversions in Article X that read:

> "In the event water from another drainage basin shall be imported into the Rio Grande basin by the United States, or Colorado or New Mexico, or any of them jointly, the State having the right to use such water shall be given proper credit therefore in the application of the schedules."

**A Few of the Messy Details Left by the 1922 Colorado River Compact**

The 1922 compact divided the water supply in the Colorado River at Lee's Ferry, near the Arizona and Utah boundary, between the upstream five and downstream three states. Conceptually, the Colorado supply was divided 50-50 based on measurements available in 1922 that projected the average annual flow in the Colorado River to be more than 15 million acre-feet per year. With this overly optimistic expectation in mind, the Upper Basin states agreed to deliver an average of 7.5 million acre-feet (MAF) a year at Lee's Ferry. On average, that would give the Upper Basin an equal amount, but that has not proven to be the case as the annual flow in the Colorado has been significantly less than originally projected. Hydrologic studies done in the 1970s indicated that the virgin flow in the Colorado at Lee's Ferry may be as much as a million acre-feet short of the 15 MAF that was the basis of the 1922 compact.

The 1922 Colorado River agreement provided that, if the U.S. entered into a future treaty with Mexico, the states would meet any treaty commitments with half the water from each group of states. As a consequence of the 1944 treaty with Mexico (requiring a delivery obligation of 1.5 MAF per year), the Upper Basin states must now release average annual flows of 8.25 MAF. Based on initial 1922 flow-assumptions, that would leave at least 6.75 MAF per year available for use in the Upper Basin, after it made deliveries to the Lower Basin states. This allocation to Mexico and the anticipation of shortages in the Upper Basin's residual allocation were some of the messy consequences of the 1922 compact.

Another issue was the 1922 Colorado River Compact language that provided that any water allocated to Native American use in any state would come from that state's share of its Colorado River water. This provision was not in keeping with Native American water rights under the *Winters Doctrine*. The doctrine came from a 1908 U.S. Supreme Court decision that ruled that when the reservations were created from the U.S. public domain, Congress had intended to also reserve any needed water supply. The Navajo's *Winters Doctrine* rights had a priority date of 1868 when the reservation was created by Congress. The only water available to satisfy the Navajo's *Winters Doctrine* demands was from the flow of the San Juan River, where the use of that water would have to come out of New Mexico's share of Colorado River water. This meant that an agreement was needed between the Navajo Nation and the State of New Mexico. The Navajo Indian Irrigation Project allocation of San Juan water would meet the Navajo's *Winters Doctrine* rights.

## 1945–1955: Early Negotiations and the Upper Basin Compact

World War II further slowed progress on New Mexico's use of San Juan River water. Planning and negotiations on New Mexico's use of its share of the Upper Colorado River water took place during the decade following the end of the war. The amount that each state would get was still in limbo at the end of WWII as an agreement among the Upper Basin states was yet to be reached. Still, several important pieces of federal legislation made it through the Congress, particularly with respect to a Navajo Indian Irrigation Project.

The yields in the Colorado River system during the 1930s and 40s gave rise to serious concerns that the Upper Basin would not have the 6.75 MAF available even in an average year for use after making annual deliveries to the downstream states. The threat of large shortages in deliveries in dry years plagued and delayed final agreements among the Upper Colorado states.

Commissioners from the Upper Basin states met in Santa Fe in 1948 where they agreed upon the division of their share of Colorado River water. The compact was ratified in 1949. New Mexico was allocated only 11.25 percent; Colorado got the biggest share, 51.75 percent, Utah 23 percent, and Wyoming 14 percent. Arizona was given an annual allotment of 50,000 acre-feet per year that was to come from the Upper Basin share. Other than Arizona, the amount that each of the other four states would receive each year was unknown as the amounts were based on how much water remained each year after deliveries of 8.25 MAF were made to the downstream states at Lee's Ferry.

Under the compact, each state would be allowed full consumptive use of its share; that means "use it all" until there is no residual water left. The 1949 compact provided a method, based on each state's percentage share of Colorado water, for determining curtailments in the event of a shortage in Lee's Ferry deliveries. The 1949 compact required that a state's share had to pay for evaporation losses from its reservoirs, for uses by the federal government, and for consumptive uses by Native Americans in each state. The allocation of funds derived from power generation, using Upper Basin water, was not addressed in the compact. Eventually, New Mexico would be assignment 17 percent of annual revenues from power generation.

Much was done during the first postwar decade, but much work remained. Massive changes had occurred on New Mexico's western slope during this period. Oil and natural gas had been found at many locations from Cuba, New Mexico westward. Good quality strippable coal was found on Navajo lands west of Farmington. Power plants were proposed and water was needed for industrial facilities. Coal-gasification was being discussed. The population in San Juan County was growing rapidly and political power was shifting. The Four Corners area was booming.

Plans for the San Juan-Chama Project proceeded, but something more important had happened by 1950. Strong, broad-based support for a Navajo Indian Irrigation Project was being voiced and planning initiated. An irrigation-water storage reservoir, to be formed by Navajo Dam on the San Juan River, was proposed and authorized by Congress. Early planning called for Navajo Dam to store 1,450,000 acre-feet of water, but an array of reservoir capacities were studied and proposed over the next twenty years.

In early 1950, the Bureau of Reclamation and the Bureau of Indian Affairs started joint work on the use of Colorado water for various New Mexico projects. San Juan-Chama water in amounts from 100,000 to 235,000 acre-feet per year were considered in early planning for these projects. While New Mexico knew how much Colorado River water it had been allocated under ideal conditions, based on the drought of the early 1950s, ideal conditions were unlikely to prevail. More importantly, the order of magnitude of shortages, and how shortages would be shared by New Mexico projects, was yet to be determined. And competition existed between potential uses.

In March 1953, Governor Ed Mechem sent a letter to the U.S. Secretary of Interior, formalizing New Mexico's allocation of water for various projects. These included: 230,000 acre-feet per year for the San Juan-Chama diversion, 23,000 acre-feet for the existing Hammond Project, and 630,000 acre-feet for the Navajo Indian Irrigation Project. The most important element in Governor Mechem's letter was his strong statement that New Mexico would support both the San Juan-Chama and the Navajo Irrigation projects, or neither project. This resolved much of the inter-basin controversy. New Mexico's desire to have just one piece of federal legislation, authorizing both projects, came as a surprise to some in Congress, but work on a single law proceeded.

Based on this letter, the Bureau of Indian Affairs generated a feasibility report in January 1955 that provided the shortages that would occur in the San Juan water supply if all of the major proposed uses were realized. This Bureau of Indian Affairs study showed that the Navajo Indian Irrigation Project would suffer an average annual shortage of 4.2 percent and a deficit of as much as 46 percent in the driest year. It was clear that some demands would have to be reduced. All uses were still up in the air and subject to competition. The Bureau of Indian Affairs report assumed that both the Navajo Indian Irrigation Project and the San Juan-Chama diversion would have equal priority on use of San Juan River water. However, because of uncertainties about ultimate requirements for both the Rio Grande and San Juan basins projects, New Mexico found that it was not desirable in the middle-1950s to make final decisions between the two basins. The diversion of western water into the Rio Grande was going to be a hard sell, particularly to the Lower Basin states. War fever was infecting all sides.

## 1955 to 1975: Twenty Years of Push and Pull

During the two decades between 1955 and 1975, a great deal was accomplished. The structural units in the San Juan-Chama Project had been built and water was being transferred across the Continental Divide. Navajo Dam had been built and the associated irrigation project was in its final planning stages. Most of the agreements between the Navajo Nation and the state had progressed to the point that the Navajo Indian Irrigation Project could begin. Some western slope industrial projects had been given water-rights. Here is how it happened.

The Opposition to Either or Both New Mexico Projects

Opposition to aspects of both the Navajo Irrigation and San Juan-Chama diversion projects developed quickly and broadly. Opposition came from California in the Lower Basin states, from some Upper Basin states, from Texas and Colorado, which were parties to the Rio Grande Compact, from other irrigation districts like the Elephant Butte Irrigation District, and from other American Indian Tribes.

And, of course, there was still opposition on the part of western slope interests to the transfer of any San Juan River water across the Continental Divide. In a 1960s speech, State Engineer Steve Reynolds told of facing opposition to an inter-basin transfer at a meeting in San Juan County: "I asked the Chairman of the Board of the Bloomfield Irrigation District: 'Wouldn't you rather see some of the San Juan River water diverted and used in the Rio Grande than see it go down the river to California and Arizona?' The Chairman replied: 'NO, and if someone is going to bed with my wife, I would rather it won't be my brother'. Such an emotional reaction to water uses, national, state, and local is not unusual." And the water war went on.

The most serious demands to stop New Mexico's use of any Colorado River water came from California. In the late 1950s, the California Colorado River Board produced an analysis of the impact of proposed developments on Colorado River water. Funds from power generation, water quality, and the economics of water use were considered. Many of the findings were critical of New Mexico's proposed projects. However, the study was biased in a number of respects: derived consumptive-use estimates did not include a reduction for precipitation; the interest rates used in the economic analysis treated the San Juan-Chama Project irrigation supply as water used on new farmlands and not as a supplemental supply meant to enhance the economies in central New Mexico; and the analysis made the error of assessing interest even though federal rules indicated that interest did not apply to this type of project. The findings criticized New Mexico's proposed use of electrical power income, when the decision to do so was the state's alone, and the report demanded that water quality not be impaired although water quality was not a criterion in the allocation of Colorado water rights among the states.

The most derogatory conclusion in the California analysis was that, "the San Juan-Chama Project would be a poor national investment." In a letter, Governor Mechem dismissed these complaints as a defensive "southern California" stance. Serious opposition from California was to take the form of resistance to congressional allocations of funds to support major projects in New Mexico. California's power in Congress made it necessary for New Mexico to take their complaints seriously and to respond with positive information. While not necessarily supportive of New Mexico, the other western states were also seeking Bureau of Reclamation funds from Congress and they needed New Mexico's help.

Farm interests in mid-western and some western states, including New Mexico, voiced opposition to the Navajo Indian Irrigation Project on the basis that it would add to existing surpluses of some farm products, that the Navajo Indian Irrigation Project would require federal subsides designed to support prices for some crops, and that the project would cause unfair competition with non-Indian farmers. Cases were made and refuted by each side of the controversy. The poverty of the Navajo people was the best argument for implementing the project.

Given the uses of the San Juan River in Colorado, that state was reluctant to support an upstream diversion of the river. Water supply studies were complicated by other western slope projects involving Colorado. The San Juan-Chama transfer of water into the Rio Grande also required negotiations with Texas and Colorado because of the Rio Grande Compact. Texas and Colorado wanted to ensure that San Juan-Chama water would not be used to satisfy any New Mexico obligation under the Rio Grande Compact. It was obvious that San Juan-Chama water would be comingled with native Rio Grande water. Stream gauging and adjudication of all Rio Grande rights were needed.

Southern New Mexico's Elephant Butte Irrigation District opposed the San Juan-Chama Project on the basis that the procedures developed to account for San Juan-Chama water were not sufficient to provide adequate protection upon encroachment of the district's water supply. The district also believed that the Rio Grande Compact could not be enforced by the federal government as the U.S. government was ruled to be an "indispensible party" to the compact in a 1957 U.S. Supreme Court decision. The 1958 decision stated that the federal government was obliged to defend American Indian water-right claims in the middle Rio Grande Basin.

Demands for water came from all fronts, even a few unexpected places like U.S. Public Health Service (PHS) demands that dilution water be provided for treated wastewater effluents that were discharged into the San Juan above the Navajo village of Shiprock. Leakage from Navajo Dam proved to be sufficient to satisfy the PHS.

There were also calls for dilution water because of environmental concerns that excessive amounts of salts would be leached into the San Juan by irrigation return-flows from Navajo lands. This problem was resolved in part by the inclusion into the Navajo Indian Irrigation Project the most suitable lands in the Shiprock and south San Juan area. Some of these better quality lands were obtained through trades with the State Land Office and with the federal Bureau of Land Management. Calculations showed that under full Navajo Indian Irrigation Project development, the increase in total dissolved solids in the Colorado River would be a little more than one percent. Negotiations on salinity issues between the Colorado River states and the U.S. Environmental Protection Agency continue today.

In early 1975, the state engineer filed a suit in New Mexico District Court seeking an adjudication of all water rights in the San Juan River system including all federal claims. Four days later, the Jicarilla Apache Tribe filed a motion in federal court seeking to intervene in an ongoing case challenging the San Juan-Chama Project, claiming that *Winters Doctrine* rights of the tribe would be impaired by the San Juan-Chama Project. An agreement between the state and the Jicarilla Apache Tribe on Tribal *Winters* rights was finally reached in 1994.

Success in the Negotiation Process under Reynolds

In his first few years as the state engineer, Steve Reynolds was faced with the daunting task of pushing both the Navajo Indian Irrigation Project and the San Juan-Chama Project. Reynolds and his gang of engineers and lawyers sharpened their negotiating and politicking skills trying to get agreements, by all sides, on how New Mexico's share of the Colorado River would be used.

In addition to the engineering work involved, Reynolds needed to initiate a broad range of administrative tasks. In June 1955, the Office of the State Engineer issued the permit to transfer Colorado River water-rights to the

Secretary of Interior for the Navajo Indian Irrigation Project, the San Juan-Chama Project, and the Hammond Irrigation Project. In May 1956, a similar transfer was made of the water that would be needed for the Animas-La Plata Project that would also require Colorado River water.

Perhaps the most important, and politically dangerous action, was Reynolds' 1956 declaration of the Middle Rio Grande Basin as an underground water basin linking the surface flow in the Rio Grande to the regional groundwater system. Albuquerque and other Middle Basin communities would be required to file an application for new wells and to off-set their eventual effects on the river by retiring existing surface rights.

The Office of the State Engineer also started planning and conducting hydrographic studies needed to adjudicate all water uses in the Rio Grande and Chama River basins above Elephant Butte Reservoir. Most of the water use was for irrigation on the many acequias north of Elephant Butte. Because of the comingling of San Juan-Chama water with native Rio Grande water that would occur, a system of determining all existing rights was needed.

New stream gauging stations were needed and funded, and measurements were made at the point of diversion on all of the region's acequias. These adjudications required mapping and measuring all the irrigated lands, determining land ownership, and mailing of tens of thousands of court notices of offers of water rights. Virtually all of the non-Indian water rights were adjudicated prior to the completion in 1971 of the structures in the San Juan-Chama Project.

Reynolds dealt forcefully with this myriad of issues including the really tough ones: (1) how much water was to be allocated (2) to what use and (3) how to reserve water for periods of shortages in the Upper Basin states' deliveries at Lee's Ferry. When reading the history of how these issues were resolved, it seems very straight forward and "no big deal" to achieve. But it was a huge deal that required constant negotiations, changes, engineering, and many calculations. It also required help from more than a few arm-twisters in the U.S. Congress: Tom Morris in the House, Chavez and Anderson in the Senate, and later Senator Mechem. Governors Mechem and Simms did their share of letter writing. Reynolds was and remained the key in the process.

The first step in the negotiations process was a series of conferences held around the West on Navajo Indian water demands. Reynolds and/or his staff attended these sessions and often took the lead in getting agreements. The federal government was represented by the Bureau of Indian Affairs and the

Bureau of Reclamation, and members of the Navajo Tribal Council attended. The first major breakthrough was at a fall 1955 session, when it was agreed that, for study purposes, only Indian irrigation projects would be discussed. The Navajo Indian Irrigation Project was to be on a non-reimbursable basis, where the tribe would not have to pay for the construction of Navajo Dam or the irrigation facilities.

Over the next twenty years, it was a step-by-step process where the amount of Colorado River water allocated to various projects was reduced to provide for periods of drought, to address municipal and industrial needs in San Juan County, and to satisfy political concerns. Even though the San Juan-Chama Project had been thought of in terms of 235,000 acre-feet per year in these years of negotiations, New Mexico agreed to limit its initial San Juan-Chama allocation to 110,000 acre-feet per year.

A succession of federal laws was enacted in efforts to settle a number of Navajo Nation issues. For example, was the Nation willing to forego all of its *Winters Doctrine* claims for the Navajo Indian Irrigation Project? Were they willing to accept a priority date of 1955 for their San Juan water rather than 1868, the date of congressional creation of their reservation? Were they willing to accept pro rata share shortages of San Juan water based on their allocation? Would they accept the use of San Juan water for industrial development in the Four Corners region? Answers to these questions in congressional testimony, in correspondence, and in meetings held by representatives of the Navajo Nation, were taken to all be positive.

The Bureau of Reclamation and the Office of the State Engineer continued to conduct engineering studies designed to strengthen the various projects. For example, a study was made to determine the feasibility of using sprinkler irrigation for the entire Navajo Indian Irrigation Project. The study found that an all-sprinkler system would be economically superior as most of the lands were sandy and rolling. The best aspect of a change to sprinkler irrigation was that much less water would be needed. As a result, the annual water requirements for Navajo Indian Irrigation Project were reduced from 508,000 acre-feet per year to 330,000 acre-feet of consumptive use per year. The irrigated acreage under Navajo Indian Irrigation Project was set at 110,630 acres.

Working with the other Upper Basin states, New Mexico's beneficial consumptive use was set at 727,000 acre-feet per year. This agreement did not address the years in which shortages in deliveries occurred to the Lower Basin states. A Department of Interior study estimated New Mexico's minimum annual allocation of Colorado River water at about 647,000 acre-feet.

New Mexico also obtained permission to use an advance on the power credit revenues from Navajo Dam and other Upper Colorado River power projects to make it possible to undertake construction of the initial phase of the San Juan-Chama Diversion Project. Priorities among uses of San Juan water in the San Juan-Chama Project would follow an order with first, municipal and industrial supplies, then water supplies for irrigation in depressed areas in northern New Mexico, and finally supplemental irrigation uses.

## The Situation at the End of 1975

At the end of twenty years of negotiations on New Mexico's allocation and use of its Colorado River water, the situation looked "rosy." The Office of the State Engineer, the Bureau of Reclamation, and New Mexico's congressional forces had all done their jobs well. The only thing left to do was to finish building the needed facilities to put the San Juan water to beneficial use.

## Present Status

On the surface, it appears that the years of work have been very successful; nearly everything is built. Navajo Indian Irrigation Project is still not finished, but the San Juan-Chama Diversion is doing well. The following section provides the status of each of the major Colorado River projects as of 2010.

The Current San Juan-Chama Project

The San Juan-Chama Project takes water from the Navajo, Little Navajo, and Blanco Rivers, the upper tributaries of the San Juan River, via a system of diversion dams and tunnels and concrete-lined channels to transport water to Azotea Creek in the Rio Grande Basin. The imported waters flow down Azotea and Willow Creeks to Heron Reservoir.

By 2010, the transmountain diversion had been in service for almost 40 years. A small amount of the 110,000 acre-feet annual transfer is assigned to evaporation losses in Heron Lake and to river-transport losses. By 2010, many of the municipalities were just beginning to use the surface water made available to them. To use 48,000 acre-feet of annual consumptive use, Albuquerque has built an elaborate diversion dam, treatment works, and transmission system. Assuming a 50 percent return flow to the Rio Grande, Albuquerque has built their facilities to divert 96,000 acre-feet per year from the river. In 2010, Santa Fe was working on plans to divert their 5,605 acre-feet annual allocation and Los Alamos County was in the process of negotiating a contract for engineering work on their 1,200 acre-feet allocation. Other communities with allocations include: Twining Water and Sanitation District, 15 acre-feet; City of Espanola, 1,000 acre-feet; Village of Taos, 400 acre-feet; Town of Belen, 500 acre-feet; Town of Bernalillo, 400 acre-feet; and the Village of Los Lunas, 400 acre-feet. Most of these communities do not have active plans for use of the water, but there have been short-term leases with downstream municipalities on the Rio Grande.

Other annual allocations of San Juan-Chama water include supplemental irrigation water to the Middle Rio Grande Conservancy District (20,900 acre-feet), the Pojoaque Valley Irrigation District, (1,030 acre-feet of which 34 percent are to Indian lands), and to Nambe Falls Reservoir (with a capacity of 2,023 acre-feet providing storage for the Pojoaque area farms). These water allocations have already been put to use. Some water has been assigned to fill conservation pools in various lakes. An annual allocation of 5,000 acre-feet is available for Cochiti Reservoir for a minimum pool of 1,200 surface acres. As a part of its settlement of its *Winters Doctrine* rights, the Jicarilla Apache Tribe has been allocated 6,500 acre-feet of San Juan-Chama water. There remains 4,990 acre-feet per year not under contract.

The Current Navajo Irrigation Project

Navajo Dam was completed in 1962 with an active storage capacity of 1,708,600 acre-feet. The reservoir extends into Colorado and provides recreational and flood control as well as its principal purpose of storing water for the Navajo Indian Irrigation Project. When, and if it is completed, the Navajo Indian Irrigation Project will irrigate 110,630 acres of land by sprinklers to grow alfalfa, corn, wheat, barley, potatoes, onions, pinto beans and pasture, benefiting some 170,000 members of the Navajo Nation. The facilities are being constructed in eleven blocks of approximately 10,000 acres each. By 2010, seven blocks, about 63,881 acres, were under irrigation. Block 8 is under construction. All other project lands are currently used for grazing.

The Status of Other Bureau of Reclamation Contracts for San Juan Water

The Colorado River water has also created municipal and industrial water uses in the San Juan Basin. Congress authorized these additional uses under separate contracts that must first be approved by Congress. Over the years, several industrial contracts for San Juan water have been approved by the Bureau of Reclamation and Congress. All have expiration dates, allocation quantities, and points and purposes of use; some contracts have expired and some specifications have changed. The largest of these industrial allocations has been to Public Service Company of New Mexico for its Four Corners coal-fired power plant and to the Utah Construction and Mining Company.

The Hammond Irrigation Project and the Animas-La Plata Project have both received San Juan water allocations from Congress. The Hammond Project is an existing Bureau of Reclamation project that diverts water directly from the San Juan River to irrigate 3,900 acres of alfalfa, wheat, barley, and pasture land. In 1994, the Bureau of Reclamation prepared an assessment of the Hammond Salinity Control Project. The Animas-La Plata Project in southwestern Colorado, currently under construction, will develop flows of the Animas and La Plata rivers. The project will provide storage of New Mexico water for municipal and industrial use and for the Navajo Nation.

In 1986, the cities of Aztec, Bloomfield, and Farmington joined San Juan County and the Rural Water Users Association to form the San Juan Water Commission. These communities take their municipal supplies from either the Animas River, and/or the San Juan River. The Commission represents the area's interests in efforts to build the Animas-La Plata Project to gain additional storage facilities

**War Drums Close at Hand?**

The future use of New Mexico's Colorado River water may not be as "rosy" as might be presumed in view of all the negotiations that have occurred over the years. In recent years, the Navajo Tribal Council has voiced concerns about the Navajo Indian Irrigation Project and its *Winters Doctrine* rights. One problem is that funding has not been provided uniformly over time for the continuous development of the Navajo Indian Irrigation Project. Congressional funding problems for the Navajo Indian Irrigation Project started in the late 1960s. Some concerns were based on the Bureau of Reclamation's questions about the lack of cost-effectiveness of transporting water to some of the more distant project lands. Delays were then encountered in converting some of the early farm-blocks into effective farms. The Tribes ability to farm the total acreage effectively was a concern. During the Reagan administration, no funding was provided for the Navajo Indian Irrigation Project. Congress resumed funding for Block 8 in 1989, which increased the farmed area to 70,000 acres.

Over the years, completion date estimates for the entire 110,630 acres have moved from the late 1970s to 1996, and now to an uncertain completion date. The project was severely under-capitalized from the start. Naturally, the Navajo Tribe has been concerned about the lack of funding and has perceived it as a deliberate effort to choke off the project.

Many of the old questions about Navajo *Winters Doctrine* rights, and the priority of other Colorado River water projects have resurfaced. What did the Navajo Tribal Council agree to in the 1950s and 60s? It appears that all that the tribe had said "yes" to was that all San Juan water uses that passed through Navajo Dam would have an equal priority with a 1955 date. Will the Tribe assert a *Winters Doctrine* claim to San Juan water "not passing through" Navajo Dam? In an unpublished file memorandum in August 1988, Reynolds reviewed the testimony of tribal representatives and the 1962 Act (PL 87-483) authorizing the Navajo Indian Irrigation Project. Reynolds found that some sections of the act were sufficiently ambiguous to allow for differences of opinion.

A more serious question relates to congressional intent. How much water had been guaranteed by the Congress for the Navajo Indian Irrigation Project? Was it the 508,000 acre-feet in the 1962 Act (PL 87-483), or was it the 330,000 acre-feet per year needed for sprinkler irrigation? More importantly, should the difference between these two quantities (178,000 acre-feet per year) be Navajo water? The Tribal Council thinks so. Officials in the Interior Department believe that any excess water should be subject to appropriation by others, once the Navajo Indian Irrigation Project needs are fulfilled. The provision for sharing of shortages of San Juan water between all Bureau of Reclamation's San Juan projects and the Navajo Indian Irrigation Project is in doubt, although many believed that the tribe had agreed to shortage sharing in the 1950s. It sounds like this is all headed for a major fight in the federal courts.

Further complicating these problems are the claims of the Jicarilla Apache Tribe. In 1992, the U.S. Congress passed the Jicarilla Settlement Act, which allows the Secretary of Interior to negotiate a final settlement of Jicarilla water claims. This Act recognizes the rights of the tribe to a total of 40,000 acre-feet per year of which 33,500 acre-feet is to come from the San Juan River. Where will this water come from? Will it be from some of the same water claimed by the Navajos? Settling all these claims should provide for interesting future battles.

# 14

# Whose Water Is It, Anyway? Anatomy of the Water War Between El Paso, Texas and New Mexico

Linda G. Harris

## The El Paso Surprise

El Paso was still hot in early September 1980. El Pasoans, trying to ignore the 90° heat, focused instead on the upcoming football game between the University of Texas at El Paso and Texas Tech University. UTEP's new quarterback boasted to reporters that, "The chance is there to surprise them. It's there." The weekend's surprise, however, would not be the game.

The first hint that something was up came Friday Sept. 5, 1980, when the *El Paso Times* announced that "Coming Sunday" the paper would run in-depth stories on El Paso's water problem and the "tangled legal web facing it and the surrounding area." Late that afternoon in Albuquerque, lawyers for El Paso quietly filed the first legal action in what would become a three-decade legal battle over rights to New Mexico's groundwater.

Some two years earlier El Paso's Public Service Board had begun secretly planning its challenge against New Mexico, hiring the Houston law firm of Vinson & Elkins to explore "possible litigation" to obtain new water supplies. (The 150-member firm included former Texas Governor John Connally and Paul Bloom, a former legal counsel to Steve Reynolds.) El Paso was prepared to take its case to the U.S. Supreme Court in order to overturn New Mexico's ban on exporting its groundwater.

El Paso's legal volley stunned New Mexico's water hierarchy. The state's export ban, enacted in 1941, had protected it from outsiders for four decades. In Santa Fe, New Mexico State Engineer Steve Reynolds, who was named as co-defendant in the suit, remained unflappable. However, he and his staff quickly set to work on a strategy to deal with El Paso's surprise move.

In Las Cruces, city officials didn't know what to think. The city, about one-tenth the size of El Paso and forty miles to its north, would be most affected if El Paso won the suit. In those first days, Las Cruces officials seemed perplexed by El Paso's lawsuit. At first, Mayor Joseph Camuñez said they were willing to "work something out" with El Paso. One city commissioner called the suit "unfortunate." Only later would come the animosity and bumper sticker protests. Ken Needham, who would soon become the city's utilities director, said he wasn't sure where El Paso wanted to drill wells, but that the city of Las Cruces was "very much concerned about it and we will be following what is going on." Elephant Butte Irrigation District, which served the region's farmers, was preoccupied with a major reorganization within the district. Its first reaction was to stop buying supplies from El Paso. Later it would be the most aggressive of all the parties involved in the legal proceedings.

As the New Mexicans began to regroup, they became more and more defensive. First because they thought El Paso's lawsuit was unfair and illegal. The move also shook their faith in the invincibility of their own water law. And they were just plain angry because in this part of the country, water meant life and livelihood. El Paso was threatening both.

El Paso too was fighting for its economic life. Its population was predicted to grow to more than a million people over the next 50 years and the city claimed it had no reliable water supply to meet that demand. Along El Paso's western edge, the Rio Grande marked the border with Mexico and was the only barrier to illegal immigrants trying to cross into the United States. In its view El Paso's population demands warranted a share of New Mexico's water. In 1986, after six years of back and forth legal maneuvers, State Engineer Steve Reynolds agreed to a hearing of El Paso's request for water. The hearing brought both sides to face each other in an airless conference room at New Mexico State University in Las Cruces. By then the suit had involved millions of dollars, dozens of lawyers, and the opinions of every man on the street and politician running for office. Reynolds, by New Mexico law, served as presiding officer. Soon after he was named state engineer in 1955, it became apparent that the new *mayordomo*, would prove to be a smart and powerful water boss. Someone once described Reynolds as a "craggy Gary Cooper." It was Reynolds who chose the hearing room, which was barely large enough to seat all the legal and technical experts, let alone many bystanders.

From November 1986 to August 1987, the proceedings consumed 54 days, 33 witnesses and 13,000 double-spaced pages of testimony. It was a crucial process for both sides. Throughout, Reynolds paid careful attention to each witness and statistic, sometimes displaying his own expertise in hydrology and water law, as well as a finely honed sense of humor. For instance, as the proceedings opened, he asked that the record show he would be maintaining order using his (pocket) knife as a gavel.

## A Grand River

The very nature of water makes it an elusive subject at best. Because it honors only geologic boundaries and the laws of hydrology, water can complicate man's most reasoned attempts at regulation. That is particularly true in this corner of the Southwest where El Paso and New Mexico share common hydrological resources. The region receives some 8 inches of precipitation a year, mostly during July and August. Early on, farmers had laid claim to the surface waters of the Rio Grande for irrigation, leaving cities like El Paso and Las Cruces forced to depend upon groundwater aquifers for their municipal needs. When the river runs low, farmers also tap into the aquifers for supplemental irrigation.

Although El Paso's legal battle was over rights to groundwater in New Mexico, the Rio Grande and its underflow dominate the overall water supply picture. The river, which begins as snowmelt, winds its way through the San Luis Valley in southern Colorado, pure and blue as the sky. By the time it reaches the border between New Mexico and Texas, the river, slowed by silt, is the color of milk chocolate. At El Paso it angles southeast until it reaches the Gulf of Mexico, a meandering 1,800-some miles from its headwaters in Colorado. While today's Rio Grande hardly lives up to its "grand river" title (the flow of the Colorado River is 17 times greater), it reins as the region's only renewable water supply.

For most of its history the region's most important river also has been most unpredictable. Historically the Rio Grande was a shallow meandering river during the winter. But heavy spring snowmelt and summer storms could quickly send the river overflowing its low banks, creating a floodplain as much as five miles wide in parts of the Mesilla Valley. During the flood of 1865, the Rio

Grande dramatically altered its course through the valley. During the flood, the river cut a new course west of Mesilla, leaving the village situated on the river's east bank. El Paso experienced a similar boundary shift in 1867 when the flooding Rio Grande changed its course and moved south into Mexico. The new channel created an island, known as the Chamizal, between the old river border and Mexico. The Chamizal boundary dispute was finally settled in 1963 when both nations signed an agreement dividing the land between them.

Every ten summers or so, the river also ran dry, with droughts sometimes lasting several months. The drought of 1879 was so severe that farmers in Las Cruces abandoned some 2,500 acres of farmland for lack of irrigation water. Demands on the river had increased so much following the Civil War that by 1880, farmers were irrigating nearly every piece of farmland along the Rio Grande in Colorado and New Mexico. By the 1890s, upstream irrigators were taking so much water from the river that farmers downstream in the Mesilla and El Paso valleys as well as in Mexico sued for a guaranteed share of the Rio Grande. In order to allocate the shares, the U.S. Bureau of Reclamation started the Rio Grande Project, which furnishes irrigation water for about 178,000 acres of land. Part of the project included Elephant Butte Dam, which was completed in 1916 to store water for downstream users. Below Elephant Butte and Caballo dams, the Rio Grande now follows a course laid out by the Bureau to control flooding. Also, because of its importance to the region's economy and the number who share its supply, the water is released from the reservoirs according to treaty obligations to Mexico, provisions of contracts entered into between the irrigation districts in the Rio Grande Project, and other legal and legislative mandates.

The Rio Grande in southern New Mexico is designated as the Lower Rio Grande (because it's in the lower part of New Mexico), but once the river crosses into Texas it is called the Upper Rio Grande. The river is differentiated this way to simplify how it is studied, regulated, and used. While farmers are guaranteed irrigation rights to the river's waters, the amount available to them varies from year to year, usually depending upon precipitation. For example, heavy snows in the winters of 1983 through 1985 increased the amount of water available for irrigation by more than six times the average for the previous forty years. Water, particularly for irrigation, is measured by the acre-foot, which is the amount of water it takes to cover one acre of land one foot deep.

The supply to the Lower Rio Grande comes from three sources: natural runoff, irrigation return flow, and treated sewage effluent. Runoff, primarily spring snowmelt, depends upon snowfall in the Colorado mountains the previous winter. Irrigation return flow is excess water that is returned to the river from surface and groundwater irrigation and from water added to cropland to flush out accumulated salts. Return flow from New Mexico farms in the Lower Rio Grande is the main source of irrigation water for farmers just below the state line in Texas, and ranges from 100,000 to 200,000 acre-feet a year. Treated sewage effluent accounts for a small percentage of total return flow.

The Upper Rio Grande in Texas reaches from the New Mexico state line to Fort Quitman, Texas, some eighty miles away. Each year that stretch of the river supplies from 100,000 to 400,000 acre-feet of water to El Paso County's irrigation district, to the city of El Paso and to several other users.

About thirty percent of this irrigation water is returned to the Rio Grande's main channel. The city of El Paso returns about fifteen percent of its supply to the river each year from treated sewage effluent.

Mexico's share of Rio Grande water is diverted at the International Dam into its main channel. However, below the dam where water is allocated to the United States, water is sometimes taken without authorization from the river's main channel and diverted to farms in Mexico. Juárez, Mexico, directly across the river from El Paso, uses its sewage effluent for irrigation.

**Nature Underground**

The history of the Mesilla and the Hueco aquifers dates back some 26 million years when the Rio Grande Rift began to form. Over this long time, the earth's crust was pulled apart, causing great blocks of the crust to sink, in some places nearly 27,000 feet. These blocks created a series of basins extending hundreds of miles. Over millions of years, the basins filled with water from prehistoric streams and from sediments eroded from nearby mountains. In turn, these sediments (porous rock and gravel) filled with runoff, thus forming basins of water underground, known as aquifers or groundwater basins.

During this period, the arid climate and geologic structure of the region prevented the surface flow from reaching a drainage outlet. The cycle of low precipitation and high evaporation tended to concentrate salts in the water, which, without an outlet, percolated into the sediments. Some 300,000 years ago when the Rio Grande became a through-flowing stream, the river opened the drainage and flushed out the saline groundwaters in the upper levels of the Mesilla aquifer. As a result, water in the upper sediments of both the Mesilla and Hueco aquifers is fresh, while the deeper sediments contain deposits of saline water.

Groundwater, unlike most surface water, moves very slowly. A clever scientist once calculated groundwater's speed as 1/70$^{th}$ a snail's pace. In geologic terms, the time it takes an amount of water to infiltrate into an aquifer is called the recharge rate. The recharge rate for aquifers in southern New Mexico is less than one inch a year. During glacial times in New Mexico, when it was cooler and wetter, water moved into the aquifer five to ten times faster.

**Aquifers at Issue**

El Paso's bid for groundwater in New Mexico focused on two distinct groundwater formations—the Mesilla Basin and the Hueco Bolson. As with the Rio Grande, these aquifers lie beneath both New Mexico and Texas. The Mesilla Basin also crosses into Mexico where it shares common hydrologic characteristics with its U.S. counterparts. The dispute in the *El Paso* case involved water from the Mesilla and Hueco aquifers beneath the New Mexico side of the state line.

Mesilla Basin groundwaters follow the general boundaries of the Rio Grande floodplain from below Caballo Dam to Las Cruces. From there the boundary shifts west of the river valley and continues south into Mexico. In New Mexico when groundwater is officially declared a "basin," it is given an administrative boundary and placed under the authority of the Office of the State Engineer.

Groundwater is the major source of municipal supplies to the region's communities, especially Las Cruces, which draws from the Mesilla Basin aquifer. However, agriculture uses more than six times as much groundwater in a typical year. Although pumping costs make it more expensive, New Mexico farmers rely on it to supplement irrigation in years when the surface water allocation is low.

The value of the Mesilla Basin is not so much in how groundwater is being used today, but in how much will be available for future uses. The west mesa portion of the aquifer, for example, is reported to contain 34 million acre-feet of fresh water, and is where El Paso proposed to drill wells. It is also where southern New Mexico farmers expect to turn for their future water supplies.

In Texas, groundwater in the Mesilla Basin lies north of the pass between the Sierra de Juárez in Mexico and the Franklin Mountains in El Paso. Because the basin's shallow groundwater is part of the Rio Grande floodplain, the quantity and quality of this water is directly affected by the river's characteristics and irrigation practices both in El Paso and on upstream.

Although nearly a quarter of the Mesilla Basin lies in Mexico, very little is known about the quantity and quality of the water supply in that area. In addition, information is scarce on the future demands of Juárez, a city of some 1 million people. Any change in the municipal water demands of a city that size would affect both the quality and quantity of groundwater in the Mesilla Basin.

The Hueco Bolson in New Mexico lies to the east of the Organ and Franklin mountains and below a surface area of about 300 square miles. Although the Hueco Bolson is technically a basin, it has retained its historic designation as a bolson, which is Spanish for "large purse." Bolson is a geologic term applied to basins in the desert Southwest that have no surface drainage outlets. The Hueco Bolson is made up of alternating layers of clay, silt, sand and gravel. The water underneath these deeper sediments generally is of poorer quality than water near the surface, especially farther from the mountains where there is no runoff. The New Mexico portion of the Hueco contains at least 6 million acre-feet of recoverable fresh water. Chaparral, a community located just across the state line northeast of El Paso, draws 2,000 acre-feet a year from the aquifer for domestic and irrigation uses.

**Westward Push: El Paso**

The technology to tap the region's vast groundwater reserves coincided with the great westward expansion. Fate Kaufman, who wrote Ben Jarrett from El Paso in 1899, promoted the city as a great place to make money and to raise a family. "You can make a fortune drilling water wells for the thirsty ranchers around here," he wrote Jarrett. Since El Paso didn't drill its first well into the Texas side of the Hueco Bolson until 1903, Jarrett became a "waterman" driving a wagon through El Paso's dusty streets selling water from 55-gallon barrels.

By the turn of the century the West had become a little less wild and the railroad had brought prosperity—and people—not only to El Paso and Las Cruces, but also to Juárez across the border. Although their populations numbered about the same in the mid-1800s, El Paso's strategic location would push its population higher and faster.

Since then the Hueco Bolson has provided an increasingly larger portion of El Paso's municipal water supply. In 1985 the Hueco in Texas contained an estimated 10 million acre-feet of water and accounted for between fifty-nine and sixty-nine percent of El Paso's municipal water supply. As a result, the water table declined sixty to eighty feet in northeast El Paso. The lower water table meant deeper wells, and higher pumping costs. Also, the deeper the well, the greater the chance of drawing salt water into the fresh water zones. In efforts to increase recharge to the aquifer in that area, El Paso injected its treated sewage effluent directly into the aquifer. Near the river, the aquifer is replenished by natural recharge, irrigation return flow and treated sewage effluent.

Predicting population growth and future water demand is a complex science involving experts in areas such as hydrology, demography, economics and water law. In the *El Paso* case they estimated future consumption based on predicted population growth and a myriad of other factors including the demand from municipalities, irrigated agriculture, manufacturing and recreation. The experts, however, were at odds over how far into the future such predictions would remain reliable. New Mexico law set a forty-year limit for water use planning for municipalities and counties, while El Paso extended the limit to 100 years.

El Paso in 1860 was still an isolated outpost with 428 residents. When the railroad arrived in 1881, El Paso began to flourish. Subsequent booms were the result of the increased importance of the military at Fort Bliss. During the Mexican Revolution, for example, some 50,000 troops were garrisoned at the fort. Later military buildups during World War I and World War II fueled both the economy and population growth. By 1950, El Paso's population had grown to 130,345. In 1980, about 20,000 military and 7,000 civilians were stationed at Fort Bliss.

In 1980 El Paso's population was nearly half a million people, working mostly in the military, manufacturing and tourism sectors. Experts predicted its population to reach 1.2 million by 2020. As El Paso's population grew, so did its per capita water consumption. Before 1940, for example, each person in El Paso used just under 100 gallons a day. By 1970 the daily per capita use had increased to 195 gallons. Some of that increase was the result of a higher economic status (more homes with flush toilets, for example), and the introduction of labor saving devices such as washing machines and dishwashers.

El Paso's other major water demand is for irrigation. The irrigation water supply is controlled by the El Paso County Water Improvement District No. 1, which has water rights under the Rio Grande Project to irrigate 69,010 acres of land. About seventy-six percent of the district's irrigation needs come from surface water, with the remainder coming from groundwater.

## Across the Border: Juárez

Juárez' population growth has paralleled that of El Paso for much of its history. During the three decades from 1940 to 1970, its population increased from 48,881 to 424,135. The 1980 population was between 800,000 and 1.2 million. Like El Paso, Juárez depends upon the Hueco Bolson for its municipal water supply. Pumping on both sides of the border has led to uncontrolled draw down and water quality problems in both cities. However, Juárez officials believe that in the future the city can tap into an ample groundwater supply in aquifers located south and west of the mountains near Juárez. These waters presumably are an extension of the Mesilla Basin. The city's high birth rate and young population point to increasing future demands on the water supply. In addition, migration from the interior of Mexico also strains the city's ability to meet these demands. However, the city's low income level and inefficient water system keep the per capita water demand at an estimated ninety-five gallons a day. Juárez mainly depends upon the Hueco Bolson and the Rio Grande for its municipal supply. Mexico's 60,000 acre-foot allotment of Rio Grande water provides irrigation water to farmlands south of Juárez.

## Into New Mexico: Las Cruces

In 1854, five years after Las Cruces was founded, only 600 people lived in that small town. While the railroad brought prosperity to Las Cruces and military activities also increased its population, its growth never matched that of El Paso. Still, by the end of World War I, its population had reached 5,000, making it southern New Mexico's largest city. In 1980, 45,086 people lived in the city with another 51,254 living in the county. Per capita water use for Las Cruces proper was about 240 gallons a day. Although El Paso and Juárez continue to dominate the population picture, Las Cruces, New Mexico's second largest city, has become a major metropolitan area in its own right.

## The Evolution of Water Laws

A nineteenth-century judge once complained that understanding the nature of groundwater was "so secret, occult and concealed" that to administer it would involve "hopeless uncertainty." Although some still would agree with him, they also would agree that groundwater is too important to cities in the Southwest to be dismissed.

Understanding the issues in the *El Paso* case requires but a brief look at the evolution of water law in New Mexico and Texas. The battle between El Paso and New Mexico emphasized the differences between the laws of the two states while underscoring their common need to secure water for the future. The communal philosophy that defined New Mexico's prehistoric culture later shaped the laws that regulate use of the state's waters. From the beginning, New Mexicans have considered water as a public resource to be governed for the benefit of its citizens.

When New Mexico became a United States Territory in 1846, the federal government followed historic precedent in recognizing New Mexico's public control over its water resources. The 1907 Water Code confirmed these earlier laws and stated that all natural surface waters within the state belong to the public and are subject to state administered appropriation for beneficial use.

New Mexico governs its water resources under the rule of prior appropriation, which means that the first person to take water and put it to beneficial use owns the rights to that water. The right includes a priority date, a specific quantity of water in that right, and a diversion point, which is the location where the water can be taken, either from the river or from a well. The right is valid as long as the right continues to be used. The appropriator, however, owns only the *right* to use the water and not the water itself.

New Mexico's prior appropriation laws first governed only surface water, primarily because at the time the laws were passed, groundwater technology was in its infancy. By 1931, however, groundwater development was booming, and the groundwater code, patterned after surface water laws, was enacted.

The 1907 Water Code also established "the right to use of water" as regulated either by permit of the territorial engineer or by court decree. (When New Mexico became a state in 1912, the title changed to state engineer.) The state engineer is responsible for overseeing the state's water resources. In water rights hearings, he is both judge and jury, deciding water rights cases based on the merits of the case and the state's water laws. However, his decision can be appealed to state and federal courts.

The state engineer has jurisdiction over all the state's surface water and over groundwater in declared groundwater basins. When groundwater development appears imminent in an area, he declares the basin under state authority. In this way, the orderly development of the water resources ensures the protection of existing water rights. Anyone wanting a water right for surface water or groundwater in a declared basin must apply to the state engineer for a permit. In 1980 New Mexico's thirty-one declared groundwater basins encompassed nearly seventy-one percent of the state. By 1991, the state engineer had jurisdiction over thirty-three basins covering ninety percent of the state.

The application for a permit requires public notice of the application and that a public hearing be held on protested applications. Those who file a protest against the permit are allowed to present their objections at a hearing. The state engineer can deny the permit if appropriated water is not available, if the new use would impair existing water rights, or if approval would be contrary to conservation of water in the state or otherwise detrimental to the public welfare.

In contrast to New Mexico, Texas has depended upon an assortment of laws to govern water use in its sprawling state. One set of laws determines ownership and control of groundwater while another set governs surface water. The evolution of water law in Texas reflects its geographic and cultural diversity. Within its borders are the wetlands of East Texas, the coastal zone along the Gulf of Mexico, and the semiarid plains of West Texas.

The Spanish explorers who settled Texas introduced the Spanish concept of community irrigation ditches. When Mexico took control of Texas, the new republic retained the Spanish laws governing the use of land and water. Land, which was plentiful, was carefully classified as irrigable or non-irrigable, then apportioned by governmental grant with or without specific right for access to water.

The Anglo-Americans who came to Texas in the early 1800s introduced the English common law system of riparian water rights. Under riparian law, a landowner who has property bordering a stream also owns the water in the stream. This deviated from Spanish water law, which required specific authorization to use water.

When the two concepts created legal tangles, Texas looked to the prior appropriation doctrine, which had been adopted in many other western states. The 1913 Texas Legislature adopted prior appropriation as the law governing the state's surface waters.

The Texas Water Code still considers surface water as the property of the state and requires permits for its use. Ownership of groundwater, however, belongs to the person who owns the land above the aquifer and may be used or sold as private property. Texas courts have adopted the common law rule that a landowner has a right to take or use all the water he can pump from beneath his land.

The practical effect of this law is that one landowner can dry up an adjoining landowner's well leaving the landowner with the dry well legally helpless. Texas courts have refused to adopt "reasonable use" with respect to groundwater. As a result, Texas groundwater law has often been called the "law of the biggest pump."

In 1985 Texas established 12 underground water and conservation districts, which gives the state limited authority over groundwater. The districts generally are allowed to regulate the conservation, protection, recharge and waste prevention of groundwater. Such regulation has become a necessity because groundwater supplies sixty percent of all the water used in the state and about eighty-five percent of all agricultural usage. El Paso elected not to become an underground water and conservation district. However, El Paso County has been established as a "critical groundwater area" by the Texas Water Commission, but later could be designated a groundwater conservation district.

El Paso's recent search for more water dates to the early 1950s when the city's Public Service Board bought 44,800 acres of land between the city limits in northeast El Paso and the New Mexico state line in order to control water rights in the Hueco Bolson. By 1980 the value of this land was as high as $60 million. When the city sells parcels of land, it retains control of the groundwater rights.

When El Paso perceived its long-term water problems as becoming critical, the city stepped up its evaluation of New Mexico as a possible source of water. In February 1980, an environmental impact statement for El Paso stated that, "the alternative of importing groundwater from New Mexico is very attractive because of the low cost and large volume of supply. . . . For planning purposes, this supply must be considered unavailable to El Paso unless and until legal proceedings determine otherwise."

Unlike in the United States, Mexico owns rights to both surface and groundwater under provisions of the Mexican constitution. Because it is a nationalized resource, the government regulates its extraction and use, and prohibits development when further development would affect public interests or existing groundwater users. Mexico also delegates control of all water uses except hydroelectricity to the Mexican Ministry of Agricultural and Hydraulic Resources.

## The Legal Path

On Sept. 5, 1980, El Paso took the first step along a legal route it hoped would lead to a plentiful and free water supply from New Mexico. On that Friday after Labor Day, the city of El Paso, through the Public Service Board, filed suit in U.S. District Court in Albuquerque against Steve Reynolds individually as the New Mexico state engineer, and also New Mexico Attorney General Jeff Bingaman and the New Mexico district attorney for Doña Ana County. The city sought to overturn New Mexico's embargo statute as violating the Commerce Clause of the U.S. Constitution. The embargo statute, enacted in 1953, prevents anyone from drilling wells in New Mexico and transporting the water outside the state.

By September 12, New Mexico State Engineer Steve Reynolds "declared" both the Mesilla Basin and the Hueco Bolson under state authority. El Paso responded quickly by filing well applications for 246,000 acre-feet a year in the Mesilla Basin and for 50,000 acre-feet a year in the Hueco Bolson.

The state engineer denied all of El Paso's applications on grounds that New Mexico statute prohibited the transfer of water across the state line. El Paso took its case before U.S. District Court Judge Howard Bratton in January 1982. In July of that year the U.S. Supreme Court ruled that groundwater was to be considered an article of commerce and as such its transfer across state lines could not be restricted. In January 1983, Bratton ruled in favor of El Paso on all accounts, writing that New Mexico's embargo violated the Commerce Clause of the U.S. Constitution because it promoted New Mexico's economic advantage.

However, Bratton acknowledged a state's obligation to its citizens, adding that "The court recognizes that the conservation and preservation of water is of the utmost importance to the citizens of New Mexico." On February 22, the New Mexico Legislature quickly responded by repealing its embargo statute and passing a new statute allowing for groundwater export under certain conditions. Under the new law, permit approval also would depend upon the availability of water in the state making the application (Texas, for example) and the effect of the application on the state where the water is being withdrawn (such as New Mexico).

Consideration of the supply in both states ensured that New Mexico's water would not be exported to a state that had an adequate supply, while leaving New Mexico unable to meet its own demands. The law also contained the clause allowing the state engineer to refuse any application that "would be contrary to the conservation of water within the state or detrimental to the public welfare of the state." Bratton ruled that while the new law was constitutional in principle, the test of its constitutionality would be in its application.

In April 1985, the New Mexico Legislature passed an additional law governing groundwater. This law limited municipalities and counties to no more water than they could use within forty years from the date of application. This limitation became known as the "forty-year rule." After six years of legal and political debate, El Paso's well applications finally reached the hearing stage on November 18, 1986. El Paso elected to test the new law only on its well applications in the Hueco Bolson.

**Whose Water is It, Anyway?**

An underlying issue in the *El Paso* case was competition between agriculture and urban uses. Irrigators in both states were concerned that El Paso, like other large southwestern cities, would look to agricultural water rights to meet their urban demands. El Paso County's irrigation district resisted El Paso's attempts to obtain more surface water rights from the district. In turn Elephant Butte Irrigation District in New Mexico refused to negotiate any out-of-court settlement with El Paso. Because the U.S. Supreme Court recently had ruled that groundwater was an interstate commodity, western states were re-examining their groundwater laws. With water now classified as a commodity, it opened the door to marketing water rights throughout the Southwest and raised the question of how groundwater commerce would be administered.

The hearing on El Paso's well applications was a statistical tug-of-war, with each side citing water use projections to prove its case. Serving as referee of sorts was the New Mexico Office of the State Engineer, which governed all water rights and the use of water in the state, including applications for new rights. El Paso both threatened and tested that control, first by challenging the New Mexico groundwater export ban, and second by having its applications put to the test under New Mexico's revised water laws. State Engineer Steve Reynolds, as the hearing officer, had the authority to approve or deny El Paso's applications.

El Paso's Public Service Board became a major player in 1980 when it challenged and defeated New Mexico's groundwater export embargo. The board, which managed El Paso Water Utilities, called New Mexico's groundwater supply its "bridge to the future." El Paso regarded the groundwater as mutually beneficial to the social and economic welfare of El Paso and southern New Mexico. John Hickerson, general manager of El Paso Water Utilities from 1965 to 1988, was one of two key witnesses at the water hearing. There, during days of testimony, the soft spoken Hickerson chronicled El Paso's search for additional water supplies. Lee Wilson, El Paso's consulting hydrologist, presented the bulk of El Paso's technical testimony. As head of a Santa Fe-based consulting firm, he had done work for several cities in New Mexico. The hearing put him in the awkward position of testifying against some of his previous New Mexico clients.

The state of New Mexico also found itself in the tricky situation of providing evidence that could prove the case against New Mexico itself. As with all applications for water rights, the state participated in the hearing through the Office of the State Engineer's Water Rights and Technical divisions, which presented technical evidence relevant to El Paso's applications. The New Mexico Attorney General's Office also presented evidence through expert witnesses.

Las Cruces entered the legal contest as a protestant against El Paso. Its citizens had launched an economic boycott against El Paso and were displaying bumper stickers that read, "Thou shalt not covet thy neighbor's water." When El Paso filed its applications in 1980, Las Cruces was forced to wait in line behind El Paso before its own application for more water could be approved.

Chaparral, New Mexico, sits atop the Hueco Bolson on the outskirts of El Paso, just inside the New Mexico state line. The community had held a rummage sale and a benefit dinner dance to help pay its $47,800 share of the legal costs for the hearing. Chaparral's population of 5,000 would be most directly affected by increased drilling in the Hueco Bolson on either side of the state line.

Other New Mexico entities, including Alamogordo, Doña Ana County and neighboring Lincoln and Otero counties also filed as protestants. While none owned water rights in the disputed area, they argued that approval of El Paso's well applications would harm the public welfare and lower both water quality and property values in their communities. New Mexico State University, as the state's publicly funded land grant college, contended that its interests directly related to the concept of public welfare. NMSU's mandate includes educating students and providing agricultural and other water-related research for the state and its citizens. New Mexico law recognizes state universities under its forty-year rule for water planning.

The 90,640-acre Elephant Butte Irrigation District represented the most vocal and resolute group in protesting El Paso's applications. Prior to the hearing, the New Mexico district steadfastly had refused to negotiate any settlement. The district also had filed a Stream Adjudication complaint to force the state engineer to conduct a water rights inventory in the Mesilla Basin. Such an inventory, and the resulting legal challenges could tie up all applications for new water uses for 15 years or more. New Mexico's largest pecan grower, Stahmann Farms Inc., spent $750,000 in its fight against El Paso. Although irrigated agriculture in the area depends primarily upon the Rio Grande, many growers supplement their surface water irrigation with groundwater. Farmers believed increased pumping in the Mesilla Basin eventually would cause a decrease in the Rio Grande supply and degrade the quality of the water.

The State Land Office in New Mexico also had an interest in the outcome. The office manages 6,267 acres of trust lands in the Hueco Bolson and 175,000 acres in the Mesilla Basin. A majority of trust lands revenue goes to support education in the state. The office said if El Paso were allowed to drill wells on state lands, it would decrease the water supply and in turn reduce income generated from those lands meant for education. Parties to an international treaty and the Rio Grande Compact also had a stake in the outcome of the El Paso suit. The 1906 treaty with the United States entitled Mexico to 60,000 acre-feet of water a year from the Rio Grande. The Rio Grande Compact of 1938 divides the river's water among Colorado, New Mexico and Texas. The compact obligates these three states to limit uses to their share of the supply. Hydrologists contended that the Mesilla Basin was hydrologically connected to the Rio Grande and that over-pumping the aquifer would in turn decrease New Mexico's portion of the river supply, which would be illegal under the compact.

Although most El Pasoans seemed pained by their city's custody battle with New Mexico, El Paso has always looked upon southern New Mexico as a geographic extension of itself. New Mexico's pecan orchards are a favorite destination for Sunday drives and the Sacramento Mountains serve as a cool retreat from El Paso's hot summers.

It followed that when El Paso's Public Service board began searching for more water to supply its growing city, it looked to New Mexico and hired experts and lawyers to bolster their claim. The board, as the applicant, bore the burden of proving it should be allowed to drill wells in New Mexico and transport that water to Texas. El Paso had to convince the state engineer that it needed the water, that it had explored sources other than those in New Mexico, and that it had been diligent in its water conservation.

El Paso favored using 100 years for long-range water planning, arguing that New Mexico's forty-year rule was impractical for such planning purposes. The court refused to hear its legal complaint until the forty-year rule had been first tested in a hearing—the hearing that soon would begin.

**El Paso Argues its Case**

So with forty years as its horizon, El Paso set out to make its case. It based its arguments on facts relating to water supply and demand in New Mexico and El Paso. El Paso also addressed several factors that the state engineer could use in determining permit approval. Five of those carried considerable weight in the approval process. They were: beneficial use, basin administration, availability of unappropriated water, impairment of existing rights, and permit conditions. The New Mexico State Legislature added three more factors in 1983: conservation, public welfare of New Mexico citizens, and the supply and demand of water in both the basin where the well would be drilled and where the water would be used.

While El Paso addressed each factor in detail, the outcome was expected to hinge on the factors of supply and demand under the forty-year rule. Both sides in the case agreed that El Paso's proposed use of New Mexico water for municipal and industrial use was beneficial. They also agreed that the state engineer should allow mining of the Hueco Bolson. Under the condition of mining, water is taken out of the groundwater supply faster than it is being replenished.

**Interstate Commerce**

El Paso argued that New Mexico couldn't deny the permits simply because New Mexico said its uses were more important that El Paso's. According to New Mexico law, the priority date of the application is more important than the proposed use as long as the use is beneficial.

Second, El Paso asserted that the New Mexico constitution prohibits the use of new legislation that would change the outcome of a pending case. El Paso first made application for Hueco water three years before the legislature added the three new factors. Third, El Paso cited Judge Bratton's ruling on the embargo, which stated, "A state may not limit water exports merely to protect local economic interests." Citing *Sporhase*, El Paso said New Mexico must consider water as an article of interstate commerce, even in the case of shortages.

## Public Welfare

El Paso said if its well applications were approved, the overall public welfare of the region also would improve. The city said that as the hub of both West Texas and Southern New Mexico, the additional water would benefit New Mexicans as well as Texans. The consideration of supply and demand required the state engineer to review a combination of factors relating to possible water shortages in New Mexico and water availability in the applicant's state. El Paso said there was no evidence that New Mexico would experience significant shortages in the foreseeable future. Citing a study by University of New Mexico law professor Charles DuMars, El Paso estimated that the state of New Mexico contained 135 to 155 million acre-feet of unappropriated groundwater. According to El Paso, this amount was 400 to 450 times the current annual use of the entire Rio Grande system in New Mexico. El Paso said even if it fully developed the proposed wells, it didn't anticipate resulting shortages in New Mexico.

Using development costs for new water supplies and continued use of existing water sources as a gauge, El Paso presented a detailed analysis of the feasibility of six water supply alternatives. According to its study, El Paso concluded that importing water from New Mexico was its most feasible choice.

## Conservation

El Paso said it had been metering its water customers since the turn of the century and that it had instituted several methods to conserve its water supply. For example, it had cut losses due to leakage in the water supply system to about ten percent. Other cities in the United States consider a fifteen to twenty percent leakage loss as acceptable.

To encourage water conservation, El Paso's Water Utilities had used pricing, regulation and education to reduce water use. El Paso said regulation and education should be recognized as playing a role in keeping El Paso's consumption rate fifteen to thirty percent below the rate it calculated for New Mexico's desert cities.

The city also had blended lower quality water from the Rio Grande, the Canutillo shallow aquifer, and the brackish portion of the Hueco Bolson with higher quality water from the Hueco. At the Fred Hervey Water Reclamation Plant in northeast El Paso, the city injected treated wastewater into the city's primary drinking water aquifer to increase the supply. At the time, El Paso was one of the few cities in the United States to use this extreme method of water replacement.

**Supply and Demand**

The city's thirty-one wells drawing water from the three productive zones of the Mesilla Basin near Canutillo, Texas, are located about fifteen miles from downtown El Paso. The city planned to expand the existing capacity from 38 million gallons a day to 70 million gallons a day. This expansion would occur if demand on the west side of El Paso required it, or if a pipeline were constructed to provide an additional supply to the city's east side. However, the city said the aquifer already might be developed at near capacity. El Paso said the problem associated with increased use of the Rio Grande included reliability, quality and protracted litigation with the river's current users.

Ninety percent of the fresh groundwater within fifty miles of El Paso is located in New Mexico. El Paso said more than 60 million acre-feet might be available in the Mesilla Basin alone. For this reason, El Paso selected New Mexico groundwater as the most efficient and cost effective alternative. El Paso said it would also be logical to expand its use of the Hueco Bolson across the state line into New Mexico. The cost of developing groundwater resources in New Mexico would be about $200 per acre-foot. If El Paso were forced to rely on sources only in Texas, the cost would range from $586 to $1,000 per acre-foot. El Paso said its rate payers probably wouldn't support the higher development costs. El Paso also preferred the New Mexico water sources because the supply was expected to last longer and be of a higher quality than the Texas alternatives.

El Paso suggested that New Mexico could grant El Paso's well permits under conditions that would protect the supply. Such conditions might include continuous monitoring of wells, specific limits on annual pumping, and the protection of other users in the bolson. El Paso emphasized that it would recognize the state engineer's authority to regulate interstate usage of water. El Paso, after presenting its case to support its applications, stepped aside for New Mexico's presentation. But for El Paso, the hearing was far from over. Its lead attorney, Peter Schenkkan, questioned every fact and figure New Mexico's witnesses presented.

## New Mexico Fights Back

By and large, people in Southern New Mexico hold a utilitarian view of El Paso. They patronize El Paso's malls, use its hospitals and fly out of its airport. But after the lawsuit was filed El Paso had become an adversary and a threat. And in 1986, El Paso found itself in New Mexico's court. Before the hearing was over, New Mexico had called 31 witnesses, including landowners, economists, hydrologists, and local, state and federal officials in its protest of El Paso's applications. New Mexico's assembly of witnesses not only reflected the diversity of those who believed they would be affected by the approval, but also the range of opinions regarding the future water needs of both El Paso and Southern New Mexico.

## The Forty-year Rule

New Mexico's parties built their legal defense around the state's 1983 export statute and the forty-year rule enacted in 1985. New Mexico believed it could prove that El Paso would not need water from New Mexico before 2020, the water planning horizon under the forty-year rule.

In her testimony for New Mexico, Helen Ingram, an economist at the University of Arizona specializing in water policy, said the forty-year limit was reasonable because it prevented speculative growth. It is the tendency of Sunbelt development, she said, to project 100-year supplies to maintain the perception that there are no water shortages in the areas under development.

In projecting El Paso's demand over the next forty years, witnesses for New Mexico provided an in-depth look at El Paso's population and employment patterns, two major determinants of water demand. New Mexico's experts said El Paso had relied on population and employment figures that over-projected growth for El Paso. They also said El Paso's statistical models did not take El Paso's unique economic and social characteristics into account. El Paso projected its growth to reach 1.2 million by 2020, while experts for the state of New Mexico projected El Paso's population for 2020 at 874,000.

New Mexico's population expert, Michael Greenwood, a consultant from the Research Triangle Institute in North Carolina, said because El Paso's industrial growth was more dependent upon government employment, its income level was lower than in the rest of Texas and was expected to remain so. He said El Paso's percentage of home owners also was lower than in other Texas cities, with both characteristics pointing to a lower per capita water demand.

Greenwood explained that when economic times outside El Paso were good, people would stay in those regions rather than migrate to El Paso. He said statistics showed that for every one percent increase in national employment, El Paso's employment decreased by a third of that.

A major factor affecting El Paso's population and employment is its proximity to Mexico. Greenwood said for every 990 jobs available in El Paso, 317 were filled by migrant workers. El Paso also had become an attractive site for "maquiladoras," companies that manufactured goods in the United States, then shipped them to Mexico for cheap assembly. However, Greenwood said the economic boost from the maquiladora industry was unlikely to continue at past rates.

**Public Welfare**

Under the 1983 export statute, New Mexico argued that conservation had taken on a new meaning. Charles Howe, a Colorado economist testifying for New Mexico, said the scarcity value of a non-renewable resource such as groundwater in an arid environment also should be considered in evaluating demands. Howe said there should be a distinction between the actual accounting costs of the water and the economic and social costs of its use. Conservation, he said, was linked to the concept of public welfare.

Helen Ingram also testified that the term beneficial use was designed to serve the public welfare. She said in earlier times the concern for public welfare emphasized immediacy, or putting water to use as quickly as possible. However, since the 1960s the emphasis on water use shifted in favor of efficiency.

Ingram's research on water policy showed that in an arid state, public welfare must be determined by whether or not water is important to the political community. "Those with strong community and social organizational ties to water distribution should have much more valid claims than those with simply an economic interest," she said. If El Paso's well applications were to be approved, she said Doña Ana County's ability to provide services would decrease because the tax base would be eroded. In addition, losing control over water meant losing the opportunity to "shape the destiny" of the area.

Under the new export law, Ingram said the state engineer would now be asked to allocate water based on the concept of scarcity, and to "look at each appropriation and ask how each serves the environment and social welfare," plus consider whether the allocation would affect the perception of political control.

Jim Baca, commissioner of public lands for the New Mexico State Land Office, also testified that protecting water resources under state trust lands would be "consistent with the conservation of water in New Mexico and consistent with the public welfare of its citizens." The federal government grants state lands in trust to generate income for the state's public schools, universities and correctional facilities. The federal government also sets strict requirements for protecting special trust lands from damage. Baca said if water were no longer available for irrigation or municipal uses, the value of that land would be impaired.

Chaparral residents provided their views of how the community's public welfare would be affected by the approval of El Paso's well applications. Margaret Lang, a Chaparral land owner, said pumping in the Hueco Bolson already was causing problems with mineral deposits on air conditioners and other appliances. Delores Wright, Chaparral realtor and president of the Hueco Bolson Water Users Association, said Chaparral's pioneering spirits bound its residents to the land even under harsh conditions. "Nobody is going to throw us off the land, but we sure can't stay on it without water," she said. Chaparral has no viable source of water other than the Hueco Bolson.

**El Paso's Alternatives**

New Mexico contended that El Paso could meet its needs from several alternative sources, but that it was seeking New Mexico's water simply because it was the cheapest source. New Mexico witnesses testified that with proper management, El Paso could continue to draw on the Hueco Bolson in Texas. Groundwater hydrologist Steven P. Larson testified that El Paso had sufficient water supplies in Texas to meet its needs to the year 2020. He also said it would be feasible to increase production from the Canutillo well field in El Paso. In 1985 the Canutillo wells supplied 20 percent of El Paso's municipal demands. New Mexico projected that in 2020, the well field would yield 30,000 acre-feet a year.

New Mexico said El Paso had failed to develop surface water supplies other than to continue its policy of leasing rights from small agricultural tracts as urbanization takes them out of irrigation. El Paso had several contracts with its irrigation district that allowed more access to Rio Grande Project surface water. New Mexico presented a report prepared by the El Paso irrigation district showing that by the year 2000, at least 12,000 acres within the irrigation district would be retired from agricultural use, with its water potentially available for municipal uses. The state said El Paso could acquire additional water through condemnation of El Paso County's irrigation district lands. El Paso had testified earlier that it didn't want to "brown out" the valley by eliminating agricultural lands.

**New Mexico's Supply**

Determination of El Paso's application depended in part on the water supply available in New Mexico. Because virtually all water rights to the Rio Grande have been appropriated, no new water rights were available from that source. The water supply in the Hueco Bolson was estimated at 15 million acre-feet, but an estimate of groundwater stored in the Mesilla Basin was incomplete. However, before the basin was declared by the state engineer, the U.S. Geological Survey estimated it held 60 million acre-feet of water. An expert for the state said 90,000 acre-feet a year could be withdrawn from the Mesilla Basin for at least forty years.

No where did the sentiment against El Paso run as high as it did in the agricultural community. Members of the Elephant Butte Irrigation District in Southern New Mexico were wary not only of El Paso but also of encroaching urbanization in general. The district's farmers, led by Elephant Butte Irrigation District Board Chairman John Salopek had sworn that El Paso would get "not one drop" of New Mexico water. To the end, the district's attorneys Steve Hubert and Steve Hernandez held fast to that mandate.

While not directly affected by El Paso's application in the Hueco Bolson, the district did have a stake in the outcome of the hearing. In addition, El Paso's applications for 266 wells in the Mesilla Basin came uncomfortably close to the well fields in the Mesilla Valley's agricultural lands. Mesilla Valley farmers depend upon well water in dry years to supplement their allocations from the river, as well as a primary source for early and late season irrigations. They contended that El Paso's proposed pumping would lower the water table, diminish stream flows and increase salinity in the area.

William Saad, Elephant Butte Irrigation District manager, testified that the district was the "agricultural stronghold of the region and the only thing that makes this district strong is the water that's available to it." The U.S. Bureau of Reclamation allocates water to the district only for agriculture and only for use within the district. Prior to the hearing the district had steadfastly refused to negotiate any out-of-court settlement with El Paso.

While El Paso and Southern New Mexico exist in the same economic environment, New Mexico considers the two areas as economically independent. Sandy Peticolas, director of planning for Doña Ana County, said the county's established growth pattern would continue. Agriculture would continue to be a leading component of land use, followed by residential development with relatively small growth in the commercial and manufacturing land use sectors. Peticolas also said Las Cruces was growing to the east and north at the rate of twenty percent a year. "If El Paso were to heavily influence the growth of Las Cruces, then the growth would be south of Las Cruces and north of El Paso." James Peach, an economist at NMSU, cited research showing that in Southern New Mexico wage and salary rates were not dependent upon El Paso. NMSU, a major Doña Ana employer, also did not depend upon El Paso for student enrollment.

In considering El Paso's application, the state engineer also looked at the potential for water shortages in New Mexico and if the water being considered for export could be used to ease such a shortage. Although New Mexico's current water demands for its agricultural and population needs were being met, New Mexico feared that if El Paso's applications were granted, development in the area would be slowed. The state also would be vulnerable to other export demands, which could create shortages.

At half past noon on August 12, 1987, the hearing ended. Not with oration, for there were no closing statements, but with a sigh. With a great hauling of boxes and handshakes all around, the hearing shut down. Then State Engineer Steve Reynolds went back to Santa Fe to make his decision.

## The Battle Ends

Two days before Christmas 1987, the New Mexico state engineer denied El Paso's well applications with the statement that ". . . no water rights in New Mexico are needed by El Paso for a water development plan or to preserve its water supply for reasonably projected additional needs within forty years from the date of the applications . . . " His declaration was simple. El Paso wouldn't get the water because it didn't need it.

In invoking the forty-year rule, Reynolds treated El Paso as the law required him to treat any municipality, inside or outside New Mexico. El Paso has consistently opposed the forty-year planning limit as too restrictive. Reynolds's evaluation of El Paso's supply and demand was far different from El Paso's view. For example, in his ruling he cited state of New Mexico estimates that El Paso's population would reach 874,000 by 2020, while El Paso estimated its population to be more than 1.2 million by then. Based on his population findings, Reynolds said El Paso's available supply in 2020 would meet its estimated demand.

Reynolds said El Paso could use water from the Hueco Bolson and the Canutillo well field, plus he was convinced that El Paso could meet its needs through existing contracts with the El Paso irrigation district which gave it access to more surface water from the Rio Grande Project. El Paso also had the power to condemn property for water system purposes. He said such supply avenues should be the first priority in El Paso's water development plans. Reynolds not only denied El Paso's Hueco Bolson well applications, he also denied its applications in the Mesilla Basin, ruling that the two sets of applications were addressed as "common issues."

What the ruling didn't contain was a single mention of New Mexico's 1983 groundwater export law. The statute concerned transporting water outside the state. By declining to address the export statute, Reynolds avoided any entanglement with the constitutional issue of interstate commerce.

By January 1988, El Paso had renewed its fight by appealing the ruling to both federal and state courts. In its appeal to the U.S. District Court, El Paso accused New Mexico of "willful disregard of the U.S. Constitution's prohibition against economic protectionism." El Paso contended that although New Mexico's new export law was not unconstitutional in itself, it was in the way Reynolds applied the new law because it discriminated against El Paso. The city estimated that this discrimination would cost it at least $1 billion, the difference between the cost of New Mexico groundwater and the next cheapest supply. El Paso also charged that the forty-year rule was a "deliberate effort to bolster the longstanding New Mexico Water Embargo."

However, El Paso's attempts to have its appeals heard in federal and state courts failed. In 1989 U.S. District Judge James Parker refused to hear the case on grounds that the case was already in state court. New Mexico District Judge Manuel Sauccdo also dismissed El Paso's appeal, this time on a technicality. Meanwhile, El Paso city officials, exhausted by the court battles, in November 1989 began making conciliatory overtures toward New Mexico. By then El Paso and Las Cruces had new mayors, the Public Service Board had been revamped and Elephant Butte Irrigation District had a new manager. At a meeting in Las

Cruces, El Paso Mayor Suzie Azar apologized "for this long litigation" and the two cities exchanged "I love Las Cruces" buttons and El Paso "Amigo Man" pins.

Both sides seemed optimistic about a settlement. The El Paso Public Service Board drafted a three-page proposal in which it offered to drop the pending appeals. The proposal also suggested that El Paso buy water from New Mexico, and design a regional water development and conservation plan that would include both sides. Elephant Butte Irrigation District called the proposal "exciting." But in January 1990, both sides were again at odds, their disagreements ranging "from A to Z," according to Joe Hanson, Public Service Board chairman. Mayor Suzie Azar said El Paso would let the lawsuit run its course. On April 25 Steve Reynolds died in office.

**Cooperation Begins**

El Paso's final appeal languished in the N.M. Court of Appeals until the two parties eventually agreed to mediation. In 1991 El Paso agreed to dismiss its appeal, and abandoned its bid for New Mexico's groundwater. That year the two sides formed the New Mexico-Texas Water Commission to work toward solving their regional water issues outside of court. For both, it meant more focus on long-range water planning. El Paso revisited the water supply alternatives its Public Service Board had presented at the hearing. Based in part on those studies, in 2006 the city drew up a Regional Water Plan for Far West Texas. The plan included conjunctive use of local surface and groundwater resources, continued conservation, expansion of reclaimed water use, and importing groundwater from areas in far West Texas. In August 2007, El Paso opened the world's largest inland desalination plant.

The water suit also was a wake-up call for New Mexico, which began water planning in 1983 in light of El Paso's legal challenge. The resulting 2003 State Water Plan laid out strategies for managing the state's water resources. The plan included traditional goals such as protecting water rights and meeting interstate compact obligations. But it also added the goals of protecting the state's cultural, environmental and economic stability and promoting cooperative strategies for meeting the state's basic water resources needs.

Population projections cited at the 1986 hearing estimated El Paso's 2020 population at 1.2 million. The most recent census put its 2010 total at 665,005, a growth rate of fifteen percent over the past decade. That figure is expected to increase when some 90,000 troops and their families are transferred to El Paso as other military bases close down. The population of Las Cruces reached 93,570 in 2010, a twenty-six percent increase in the decade. In 1980, its population was 45,086.

While there were no winners in the water war between El Paso and New Mexico, neither were there any losers. Instead, both sides faced the hard fact that protecting and preserving water resources in the Southwest requires cooperation and planning on everyone's part—no matter whose water it is.

# 15

# The Mother of All Water Rights Adjudications?

## Kay Matthews

The adjudication of Lower Rio Grande water rights, currently being heard in the Third Judicial District Court in Doña Ana County, could result in a profound change in the way the State of New Mexico manages the Rio Grande. The court has ordered the state and the U.S. Government to reach settlement on the issue of the Rio Grande Project, or Elephant Butte Irrigation District (EBID), as to the amount of water rights the federal government actually owns. Scott Boyd, whose great-grandfather Nathan Boyd originally owned the Rio Grande Project, is also a party in the case along with Doña Ana County farmers, El Paso County Water Improvement District No. 1 (EPCWID#1), New Mexico State University (NMSU), and many others.

One hundred years ago the federal government seized Nathan Boyd's Rio Grande Dam and Irrigation Company. This included all rights-of-way through public lands, a diversion dam at Ft. Selden to serve the irrigation needs of Lower Rio Grande irrigators who had turned their water rights over to the company, and a dam and reservoir at Elephant Butte. It took this action under what is referred to as Application 8 citing the War Powers Act to make the claim that the Rio Grande was a navigable river above El Paso and Boyd's dam interfered with ship travel. This seizure allowed the Bureau of Reclamation to proceed with construction of Elephant Butte Dam and the administration of Lower Rio Grande water rights. Those water rights, as stated in Application 8, were "all unappropriated water of the Rio Grande," obviously not including senior, or prior appropriation water rights. After the takeover of Boyd's company, the state filed an application so that the Office of the State Engineer (OSE) could begin permitting these water rights, essentially commingling these senior, and subsequent junior, water rights. The OSE set a universal senior priority date of 1906, the date of Application 8.

Scott Boyd, who represents the Boyd estate, believes the federal government illegally acquired Rio Grande Project water rights and that the adjudication of these federal rights may now involve the determination of prior appropriation rights and how they will be managed by the OSE. Any changes in priority dates with regard to these rights could potentially impact both water rights allocations and transfers throughout the Rio Grande. Boyd has a vision and a mission to make the Rio Grande one river, with a direct flow system that protects farmers and pueblos, acequias, and wetlands. He would establish a water bank and water council to handle pre-1906 claims and cities would have to buy their water: no more water transfers, no more corruption, no more control of water rights by state and federal bureaucrats.

In order to understand how Nathan Boyd's dam projects affect the current adjudication of the Lower Rio Grande, they need to be put in historical context. In the late 1880s, before New Mexico became a state, Mexico, and El Paso area farmers were jockeying for control of the Rio Grande. Investors were looking at building reservoirs on the river because of a lack of sustainable water supply in the Mesilla Valley and El Paso-Juárez Valley. Texas was concerned about getting its allotted supply, and in 1890 the U.S. and Mexico signed an agreement that outlined guidelines for equitable river management. El Paso wanted a dam built three miles above the city (the U.S. Geological Survey, under John Wesley Powell, initially endorsed this project), but that would have flooded Mesilla Valley farmlands. When Boyd's Rio Grande Dam and Irrigation Company got the right of way to build a dam and reservoir at Elephant Butte and an irrigation dam at Ft. Selden, Juárez protested, claiming it would prevent the building of a dam closer to El Paso and Juarez, thereby robbing citizens of their international waters. While this protest was working its way through the courts, another tactic was used to fight Rio Grande Dam and Irrigation Company's projects. The U.S. Secretary of State, under the auspices of the War Powers Act and under pressure from Texas and Mexican speculators (and according to some accounts, President Teddy Roosevelt) who wanted to control the fertile agricultural lands in the area, declared the Rio Grande a "navigable river." An injunction was imposed, stopping the Rio Grande Company's work. The designation was made in spite of the fact that the U.S. Army Corps of Engineers had declared that the Rio Grande was only irrigable, not navigable. This case also worked its way through the court system, twice reaching the U.S. Supreme Court.

Initially, the New Mexico Territorial Government defended the Rio Grande Company because it saw the issue as a power struggle between Texas, Mexico, and the federal government, and its position was to protect water for New Mexicans. Nathan Boyd, as head of the company, had already agreed that in return for an annual payment of $225,000 over 20 years his company would deliver water to Mexican irrigators for $1.50 per acre-foot. Boyd, who was a medical practitioner before he married into money and became an investor, had gotten bankers from London to finance the Rio Grande Dam and Irrigation Company. He had initially been brought to New Mexico by the infamous law officer Pat Garrett to look at the possibility of a dam on the Pecos River. When that project didn't work out, he was contacted by farmers from Las Cruces. Whether his motives were altruistic, as his great-grandson Scott Boyd believes, or whether he just saw the dam projects as a good investment, he acquired 40,000 acres of Mesilla Valley land and as much as two-thirds of the bordering mesa land at a very small cost. Many farmers conveyed one half of their land for water rights to the other half. The plan was also to supply cities, industries, and other uses.

The Territory's defense of the company evaporated once the U.S. Reclamation Act was passed and the Bureau of Reclamation recommended building a dam at Elephant Butte, below the Rio Grande Company's site, with sufficient water for southern New Mexico and the El Paso-Juarez Valley. The government demanded forfeiture of the company's franchise in 1903 because of non-use of water rights, which Boyd declared was due to the injunctions imposed because of the lawsuits (by the city of Juárez and the U.S. Secretary of State). The Bureau of Reclamation built Elephant Butte Dam, its first major project, and established the Elephant Butte Irrigation District, which administers Lower Rio Grande water rights. Nathan Boyd fought this takeover for 30 years, and now his great-grandson Scott is a party to the current Lower Rio Grande adjudication that has been dragging on for ten years, but will finally determine the federal government's water rights. In late December 2009, Judge Jerald Valentine issued an order for the state of New Mexico and the federal government to "present to the Court a proposed subfile order outlining what rights they assert the United States holds in New Mexico for the United States Bureau of Reclamation Rio Grande Project" by April 8, 2010.

A quiet title suit initiated by the United States in 1997 triggered this long, complicated adjudication that is now in Judge Valentine's district court. At that time the federal government sued in federal court asking for title to virtually all the waters of the Lower Rio Grande, up to and including 2,638,860 acre-feet of storage water, to meet various uses: irrigation, municipal, and industrial within the project boundaries in New Mexico and Texas; water delivery to Mexico as stipulated in a 1906 Treaty between the U.S. and Mexico; and flood control, hydroelectric power generation, and water storage in Elephant Butte and Caballo reservoirs to meet the terms of the Rio Grande Compact (which determines water deliveries to Texas). Opposing parties to the suit were successful in convincing the federal court to abstain from hearing the case and moving it to state district court as an adjudication. The 10[th] Circuit Court of Appeals upheld that ruling.

In April of 2000, New Mexico filed a Notice of Intent to Make an Offer of Judgment Describing the United States' Storage and Diversion Rights and Proprietary Claims in the Rio Grande Project. The U.S. District Court, however, stayed the suit until the federal governments' water rights could be determined, after being assured that these rights could be "expeditiously" adjudicated in state court. Nine years later the state has yet to present an offer of judgment regarding the Elephant Butte water rights.

The case finally came to district court, which ordered that the federal government and the state come to a settlement by April of 2010. But the state issued a status report citing that it "cannot predict how much time may be required for future negotiations on the issue." Meanwhile, the OSE continues to issue thousands of offers of judgment in water rights adjudications based on the Application 8 priority date.

Scott Boyd's motion in court asks that the case move forward and the federal government's water rights be adjudicated, rights that he, of course, believes are fraudulent. According to the motion, "Further delay adjudicating the rights of the United States in this proceeding can only compound what may prove to be a monumental waste of judicial resources."

If the federal government and the state cannot reach a settlement, the court could rule on the feds' rights, which could affect water rights in the entire Rio Grande basin, as there were no declared basins in 1906 and therefore no boundaries between the lower, middle, and upper basins. Conceivably, if it is determined that the federal government's water rights are nonexistent, the court could proceed with a standard adjudication of senior water rights, invalidating the state's permits for the entire Rio Grande.

While this seems a very unlikely scenario, Rebecca Miller, one of the Mesilla Valley farmers involved in the suit, believes that "With or without a settlement the U.S. Government will have to reveal what water rights they actually own. EBID originally took our water rights and continues to interfere with our ability to adjudicate our water rights under the priority doctrine."

There are five Stream System Issues being heard in Valentine's court. These are issues that affect the interests of all or most of the parties to the adjudication. The first three issues deal with Consumptive Irrigation Requirements/Farm Delivery Requirements, EBID groundwater claims, and domestic wells. The fourth issue, Stream System Issue SS 104, gets at the heart of Scott Boyd's claim that the federal government illegally seized his great-grandfather's Rio Grande Company. Under SS 104, the determination of federal groundwater rights that were part of the Rio Grande Project will be connected to surface water rights as well. The most recently added issue, SS 105, addresses the water right claims of the Estate of Nathan Boyd, as well as any related claims by Scott Boyd as an *inter se* proceeding.

The April 8, 2010 hearing was supposedly the deadline for the federal government and the state to delineate their water rights claims. Instead, lawyers for the governmental entities involved—Elephant Butte Irrigation District, El Paso County Water Improvement District No. 1 (EPCWID#1), New Mexico, and the United States, who wanted to avoid dealing with SS 104— proposed mediation among themselves, excluding Boyd's estate and the Lower Rio Grande farmers who have subsequently joined the suit. The Court approved the mediation request and continued a stay of trial but imposed stringent time limits for a declaration of water rights. Scott Boyd and some of the Lower Rio Grande farmers have been negotiating with a natural resources law firm to represent them in this case. Their previous attorney was allowed only to argue for their motion to force the federal government to prove its claim. Now they look forward to the opportunity to prove their claims in court.

A joint motion to lift the stay was subsequently filed by four southern New Mexico entities: Stahmanns, Inc. (a large pecan farm and business in the Mesilla Valley); NMSU; New Mexico Pecan Growers; and the City of Las Cruces. Referred to as "movants," they claim that contrary to claims before the court that the parties currently in mediation are the only users of surface water and groundwater south of Elephant Butte Reservoir, they, too, have established valid water rights. Therefore, they have a direct interest in the disposition of SS 104 and request that the court allow them to participate in the determination of water rights in this adjudication. They also objected to the participation of EPCWID#1, which is not a water user, while they, as water rights owners, are prohibited from participating: "Current mediation efforts will not be hampered by lifting the stay of Issue 104. If any inconvenience results to the mediating parties, such inconvenience will not outweigh the Movants' interests in a fair opportunity to protect their claims to the groundwater resources of the Lower Rio Grande basin."

The state promptly responded to the southern New Mexico entities, complaining that there are now so many parties to the litigation—32 water right claimants, one *amicus curiae*, two intervenors, one proposed intervenor, and these four groups—that "efforts to facilitate settlement of issues between 35 participating parties [or more] would be cumbersome and complex, with a limited likelihood of success." The state also reminded the court that the "sole purpose of the mediation is to address the United States' claims," that is, whether the United States legally established any rights deriving from the Rio Grande Project.

Then, in July, 2010, things took an interesting turn. It seems that the federal government, which is being asked to prove its water rights associated with the Elephant Butte Irrigation District (EBID), neglected to file a valid Application 8, or Permit 8, with the Territory of New Mexico, in 1906, to acquire and operate the Rio Grande Project. The OSE came up empty handed after being ordered by the court to produce the Permit 8. Whether the absence of a permit will be viewed as a mere technicality or a substantive issue remains to be seen, but the court will move forward with hearings regarding the stream issues. At the August status conference regarding the adjudication, the court partially lifted its stay of trial (while the governmental parties are involved in mediation) so that other parties to the case can file a claim of interest and their legal position regarding SS 104 before the end of August. The parties in mediation could then file their objections to these senior water rights claims before September 10.

The water lawyers involved in this Lower Rio Grande adjudication are the who's who of the water world and any equitable settlement of this case will require a Herculean effort to rise above special interests. The Stein & Brockmann firm is representing the City of Las Cruces and John Utton of Sheehan, Sheehan & Stelzner is representing NMSU. Stein & Brockmann represented Española businessman Richard Cook in his challenge to the 2003 state statute that allows acequia commissioners to deny water transfers from their ditches if the transfer would be detrimental to the functioning of the acequia. Utton has long represented Santa Fe County in the Aamodt negotiations and the subsequent settlement. Other Aamodt attorneys also represent some of the Lower Rio Grande parties. These water attorneys belong to a cohort of law firms who represent entities that regard water rights as private property rights, not a public resource that can be regulated for the public good, and that economic development has precedence over other water management concerns. They are the ones who are directing water settlements all over the state that are contingent upon the movement of water from its area of origin to facilitate growth and development. As Lynn Montgomery, a mayordomo in Placitas who has been fighting to protect his acequia from development interests for many years, said in an e-mail regarding the Lower Rio Grande adjudication: "This case is definitely going to determine if we will have a priority system or we are going to depend on a hodgepodge of conflicting settlements that cause chaos and will make our rights worthless. . . . At the very least, we should be working and aligning with other senior rights holders, especially Treaty of Guadalupe Hidalgo protected rights, which include the Pueblos, . . . if we are to keep our ways and local acequia agriculture."

Scott Boyd recently sent a letter to New Mexico state legislators outlining his position and laying out his vision for management of the Rio Grande. Boyd points out that, by law, water is managed under the prior appropriation doctrine that recognizes local community and private rights of water supplies. In reality, however, the OSE, under Application 8, began to issue permits and to adjudicate preexisting water rights.

In his letter Boyd states: "Today, the question to be asked is why Application 8 remains the basis of both federal and state governments' claims. The two legal systems cannot coexist much longer, indeed they are on a collision course; and unfortunately, it is the state that is liable and will be seen as rogue. As the recent and numerous SE [State Engineer] defeats in various courts suggest, the costs of these water management and adjudication problems will increase exponentially so long as the SE's role includes settlements that replace adjudication under state laws. . . . It is the courts' exclusive jurisdiction to adjudicate both pre- and post-1906 rights that will bring about legal adjudication."

He goes on to predict that if the court concludes that neither the federal government, Texas, nor the state engineer had rights to the water to begin with, then the time is ripe to "establish a new era of legal water transfers." He proposes that the state legislature form a "Rio Grande Water Authority, a state and private water utility that would foster a new age of accountability, management, and water conservation through efficiency and sound legal water transfers. Rather than dividing the river through compact claims and uncertainty over who owns what water, the authority would build cooperation on the already developing links between all water right owners and water users in the various regions, as well as the cities and government." A "Rio Grande Council" would help build cooperation and investigate, identify, and determine all pre-1906 rights of in stream surface diversions. A strong acequia system under local rule would protect the environment and subsurface water tables for everyone and most importantly, provide food security: "Modernity has in no way changed the fact that the stability of society is contingent on its food base, and the assumption that global trade negates the state's responsibility to protect and insure so fundamental a need as food and water is false. There is a reason agriculture has the historic water rights, and a reason for the priority doctrine that protects them that goes to socioeconomic sustainability."

This may all sound like pie-in-the-sky rhetoric, but in many ways it's a distillation of the thinking by many acequia and agriculture advocates who have long opposed many of the state-proposed water settlements, particularly the Aamodt case, and who have long objected to the many water transfers that take water out of its area of origin, primarily in rural, agricultural areas, and move it to our growing cities, which own junior water rights. So giving Boyd some attention, which the mainstream media has failed to do, is also giving voice to many, who also view the state's management of our water as inequitable and unsustainable.

As the deadline for this chapter approached, Scott Boyd told me that he was prepared to sue the State of New Mexico for $100 million based on information he has obtained that the notorious Santa Fe Ring, the cohort of late 19[th] century lawyers and politicians who were responsible for the territorial land grab in New Mexico that resulted in the loss of Mexican and Spanish land grants, was also behind the forfeiture claims against his great-grandfather Nathan Boyd's Rio Grande Project. Boyd believes that it is past time for the court to rule that without Application 8, the federal government does not have valid water rights, and that it is time to lift the stay of trial and dispense with mediation because the federal government has no water rights to adjudicate.

# 16

# Future Water Wars in New Mexico

## M. Karl Wood

**Introduction**

Em Hall, noted writer and emeritus professor of land and water law at the University of New Mexico, was asked what future water wars will be like. He said that they will be the continuation of all the past wars. There are some precedents that would lead one to believe this is true. The Mt. Vernon Compact of 1785 is the oldest water compact in the United States. It resulted from a conference of delegates from Virginia and Maryland at George Washington's home. It dealt with issues of commerce, fishing, and navigation in the waters of the Potomac and Pocomoke Rivers and the Chesapeake Bay. The problems addressed then are far from being completely settled today as human populations and their priorities change.

Similarly, can we believe that settlements and agreements of past water wars in New Mexico are now final? Hall is probably right that old wars will continue and new ones can be expected.

As an example: on February 14, 2008 Elephant Butte Irrigation District and El Paso Water Improvement District No. 1 in conjunction with the Bureau of Reclamation reached a historic operating agreement that finally defined the basis for allocating Rio Grande Project Water between the New Mexico and Texas irrigation districts. This came after 30 years of discussions and negotiations. The signers joked about the new love affair starting on Valentine's Day. Just as World War I was meant and claimed to be the war to end all military wars, this agreement was meant and claimed to end all water wars between New Mexico and Texas on the Rio Grande. As a result of the agreement, the roles of the Texas and New Mexico state engineers and the role of New Mexico's Active

Water Resource Management program below Elephant Butte Dam would be minimal.

After the signing ceremony, New Mexico Water Resources Research Institute's associate director and longtime water expert, Bobby J. Creel, remarked that he thought the peace would last six months to a year. After that time, consultants and lawyers would start missing their checks from the irrigation districts and other interested parties, and they would start stirring the pot by looking for new problems to solve. In early spring 2011, the New Mexico attorney general filed a federal suit involving the operating agreement.

With a growing human population that is more and more affected by extreme climatic events, existing water wars are expected to continue into the future with new ones arising. Plus perceived centuries-old inequalities still exist. Examples go back to at least the Kearny Code of 1846 and the Treaty of Guadalupe Hidalgo in 1848. Because of unsettlement following this code and treaty, a senator from Texas wanted the U.S. government to force Mexico to take back New Mexico and Arizona. And Manual Armijo, the first Territorial Governor of New Mexico, declared, "Poor New Mexico, So far from Heaven, So close to Texas." Similar feelings toward neighboring states, the federal government, and entities within New Mexico's own borders still persist today.

What follows in this chapter is a look at several other situations around New Mexico, all tied to historic water disputes, where we might expect to see the crossed sabers and skirmishes of water wars to appear in the future.

**Pecos River**

Texas insists that New Mexico must deliver water to meet their Pecos River delivery obligations. They don't want money for compensation. The state engineer tries every year to avoid a priority call that would take water from junior water rights holders like the City of Roswell and send it on to Texas. Such an action would probably be futile and not add any more water to the river. The state engineer purchases water rights for tens of millions of dollars each year from Carlsbad area farmers to ensure delivery at the New Mexico-Texas state line. This is controversial because all tax payers in New Mexico must pay for the benefit of a few citizens along the Pecos River. It is felt that they should pay their own way or a priority call should be made. In addition, Texas may demand higher quality water although it is widely known that much of the salt in Pecos River water is picked up from natural sources in the Malagro Bend as the water enters Texas.

The Pecos River also has a huge salt cedar infestation. State and federal agencies have poured millions of dollars into control since the year 2000. No one can document one drop of water being added to the river as a result. Former Carlsbad Irrigation Manager Tom Davis pointed out that, because of cost restrictions, control was only along a narrow band adjacent to the river and not out in the floodplain. He claimed that the salt cedars served a purpose of keeping the river narrow and deep for efficient transport of water from northern reservoirs. Removing the salt cedars along the stream would result in its broadening, which would lead to increased carriage loss. This is bound to be an issue in the future as more money is spent to control salt cedar and treated areas are re-infested.

**Lower Rio Grande**

It appears in water meetings in this region that local water engineers, hydrologists, economists, and lawyers are as well informed and competent as those from state agencies in Santa Fe. The major Lower Rio Grande water providers are linked in an organization called the Lower Rio Grande Water Users Organization. One of the group's functions is to serve as a watchdog on dealings in this region by outside interests. The organization is always suspicious of divide-and-conquer tactics, such as the state engineer making generous offerings-of-judgment to the City of Las Cruces and New Mexico State University in the adjudication before offerings to Elephant Butte Irrigation District are completed. Other suspicious tactics include offerings-of-judgment to pecan growers that are much greater per acre of farmland than to farmers of other crops. This creates great consternation within the region.

Some of the climate change models predict long and severe droughts for New Mexico. With or without human influence, droughts invariably will happen, always sooner than we would like. Drought could result in a priority call on the Lower Rio Grande that could cause turmoil, as the junior water rights holders are often the most affluent, which means they can hire the lawyers. The 1906 treaty with Mexico contains a provision where Mexico's allotment is reduced in times of extraordinary drought. This term "extraordinary drought" has never been defined. It could bring pressure to adjust the treaty. Long and severe drought could also make it difficult to deliver water to Texas, as groundwater pumping could severely affect the flow of the river. And finally, it is not known what effects the new pumping of the Conejos-Medanos aquifer (southern portion of the Mesilla bolson) by the City of Juárez will have on the rest of the Mesilla bolson. This could cause international problems. Battles between the region and Texas, Mexico, state agencies in Santa Fe, and among the providers within the region are expected to continue.

## Middle Rio Grande

The Middle Rio Grande in New Mexico extends from Cochiti Dam south to about San Marcial. It comprises a complex mixture of municipalities, villages, acequias, tribes, farms, state and federal lands, and military installations, plus a few endangered species. This is probably one of the last places in New Mexico where the adjudication process will take place because it does not border any adjacent state, and because it is so complex. It bothers many water providers in this region, as it does water providers north of here, to watch Rio Grande water flow by to meet interstate compact and treaty obligations in the Lower Rio Grande region. Attempts to keep water upstream are expected. Like many places, conflicts here have been between residential users and irrigators and will probably continue into the future. More irrigated lands will need to be retired to offset the effects of groundwater pumping on the Rio Grande. Some farmers won't sell their rights at any price. In 2009, the New Mexico Legislature passed a bill that exempted from municipal condemnation any "water sources used by, water stored for use by or water rights owned or served by an acequia, community ditch, irrigation district, conservancy district or political subdivision of the state."

Bill Hume, a former policy analyst for Governor Bill Richardson appraised the new law in this way:

> "Are we likely to curtail or shut down our cities if they fail to meet offset requirements? Probably not. Could the cost of water for domestic and industrial use soar to unrealistic heights? Possibly, in which case urban growth might be foreclosed. Power of condemnation or not, the lure of high water rights prices to take water from agriculture is obvious. Can we just miss our delivery requirements under the compact as a long-term solution? In a word, no."

> "The Rio Grande is the only river that
> I have ever seen that needed irrigation."
> —Will Rogers

**Colorado River**

The Colorado River Compact has become the focus of sharp criticism, in the wake of a protracted decrease in rainfall in the region in recent years. In December 2007, a set of interim guidelines on how to allocate Colorado River water in the event of shortages was signed by the secretary of the interior. The guidelines are described as interim because they extend through 2026 and were intended to allow the system operators to gain experience with low-reservoir conditions, while the effect of climate change on the Colorado River's flow undergoes further evaluation. In 2008, Arizona Senator John McCain called for the compact to be renegotiated. Due to the Senator's seniority in the U.S. Congress and position as a presidential candidate in the 2008 election, criticism of the compact may have gained national significance.

The Colorado River is managed and operated under numerous compacts, federal laws, court decisions and decrees, contracts, and regulatory guidelines collectively known as "The Law of the River." History demonstrates that the compact and the Law of the River must be flexible in meeting new circumstances. Future situations no doubt will arise to further challenge the legal and institutional arrangements regulating the Colorado River. A contingency that would affect river management is severe sustained drought. Water shortages were not on the minds of compact negotiators; in fact, they seemed to believe that surpluses were more likely. As a result, the compact does not include provisions to deal with shortages due to drought. A prolonged drought would strain the entire system. Who then has priority water rights from a drought-stricken Colorado River? This is an ongoing debate. Reference to the compact and key elements of the Law of the River suggest some answers. Interpretations, however, vary; a different legal view might find fault with several premises, including treaty obligations to Mexico, perfected rights pre-dating the compact, tribal reserved rights, and Upper Basin-Lower Basin division.

The Navajo Nation Water Rights Settlement of 2009 includes the Navajo tribe, the federal government, and state government. It is not perfect and not well understood. The non-tribal and non-state and federal entities in the region have needs and there are misunderstandings that must be sorted out. This could take years or decades. Also there is a perceived surplus of water in the San Juan River Basin, which leads to the desire for more San Juan-Chama diversion. In recent decades, flows of the San Juan River have increased due to failing irrigation infrastructures in Colorado. These are now being replaced, which reduces San Juan River flow. Downstream users must adjust to the reductions. A potential future water fight involves the continued conflict between endangered species and hydroelectric generation. The currently listed species may become delisted, but other species may take their place.

> "Still waters run no mills."
> —Aglionby in *Life of Bickerstaff*

**Acequias**

The New Mexico Acequia Association has many concerns about what it considers threats to acequia waters and land-based livelihoods. First, unprecedented growth and development in New Mexico are driving demands to move water rights out of agriculture to urban, resort, and commercial development. According to studies of future supply and demand, acequia communities are projected to lose 30 percent to 60 percent of their water rights base and farmland to development in the next 40 years. These communities feel that acequias in areas with high water demands may be driven to extinction by water transfers because of reduced pressure head at the point of diversion and fewer families to contribute to the maintenance and governance of the acequias.

Many acequia and rural agricultural communities are economically disadvantaged and may experience a net loss of water rights as wealthier individuals, entities, and regions acquire water rights from a position of greater economic power. Demands to move water out of acequias come at a time when these communities are dedicating themselves to revitalizing agriculture and rebuilding local food systems. Erosion of the acequia water rights base could foreclose future options for rural community development. Further, these rural water supplies are threatened by various sources of water contamination including mining runoff, lack of wastewater treatment facilities, improper dumping of solid, chemical, and radioactive waste, and urban drainage. Traditional environmental knowledge embodied in the acequia culture is at risk because of a lack of intentional efforts in our educational systems to recognize its importance and incorporate it into curricula. And lastly, the poor condition of surrounding watersheds—from overstocked forests and invasive species—is likely to reduce stream flows in rivers, which impact wildlife and water quality.

The New Mexico Acequia Association has a plan of action.[1] The plan is to cultivate acequia lands with ancestral crops, using native seeds, and continuously improve farm and ranch soils to enhance efficient use of water. Members will actively participate in the governance of acequias and encourage new leaders to serve as commissioners and mayordomos. They will continue to celebrate their culture through funciónes, cambalaches, and festivales that honor traditional feast days and the culturally and spiritually important days in the growing season. Acequia members are seeking mutual solutions to meet local water rights needs by supporting collaboration between acequias and domestic water consumer associations in securing safe and healthy water for families.

The association is working to retain local ownership and control of water rights by strengthening acequia governance and preventing the transfer of acequia water rights out of the respective communities and the basins where they have historically existed. Association members are establishing projects to strengthen farms and ranches as part of their way of life and as part of their livelihoods. Appropriate resources are sought to improve food system infrastructure locally and regionally. Members establish community-based processes and centers for the documentation of indigenous and traditional environmental knowledge about watersheds, acequia traditions, agricultural practices, and food traditions. And finally, members are actively challenging the economic and political forces in New Mexico that result in growth and development patterns that are transforming the landscape and undermining their way of life.

Can the acequia associations make all of this happen? It is not likely in the near future. Nor is it likely without challenges within acequias, between acequias, or between acequias and non-acequia interests.

**Federal Power and Reserved Rights**

Opening interstate compacts and international treaties can be dangerous. Of course everyone wants more of the pie without giving away anything substantial. Texas, Chihuahua, and New Mexico may not like everything in the 1906 treaty, which is restricted to the Paso del Norte region. But there are always fears by those directly affected by the 1906 treaty that the 1944 treaty with Mexico could be opened for renegotiation. The waters of the Paso del

Norte Region could become trading chips for new agreements on the Colorado River and the Lower Rio Grande if the 1906 treaty is opened also. Similarly, opening the Pecos River Compact between Texas and New Mexico could lead to Texas insisting on opening the Rio Grande Compact at the same time. New Mexico probably has more to lose on the Rio Grande than it has to gain on the Pecos River.

In 2003, Senator Jeff Bingaman of New Mexico held a U.S. Senate Field Hearing on water at New Mexico State University. Bingaman concluded from the hearing that not enough was known about the transboundary aquifers along the U.S.-Mexican border. He sponsored, and the U.S. Congress passed, a Transboundary Aquifer Assessment Act, which was signed by the president in December 2006. The act was delayed at least a year and almost indefinitely because of a state-federal government dispute. The act was written to characterize the transboundary aquifers as to their location, depth, thickness, flow direction, flow rate, quality, and other factors. No recommendations for management would come from the act. The act specified that the program would be administered by the Water Resources Research Institutes and U.S. Geological Survey state offices in New Mexico, Texas, and Arizona. The U.S. International Boundary and Water Commission declared itself to be a major player. To get states' support for passage of the act, meetings were held with the state engineers in Arizona, to be followed by meetings with New Mexico and Texas.

The Arizona meeting was attended by the director of the Arizona Water Resources Research Institute, representatives from the U.S. Geological Survey, and the Commissioner of the International Boundary and Water Commission (IBWC). In the meeting, the IBWC stated that it would use the information from the Act to write a groundwater treaty with Mexico. Personnel at the meeting from the Arizona state engineer's office were shocked. They saw a new treaty as a threat to state rights and an attempt to take ownership of Arizona's water. They warned the New Mexico state engineer of IBWC's intended visit and agenda. In a subsequent meeting with New Mexico's state engineer, IBWC representatives again stated their intentions of using information from the Act to write a groundwater treaty. New Mexico State Engineer Tom Turney replied simply, "The hell you say." The state engineers from Arizona and New Mexico in turn warned the Texas Water Development Board of IBWC's impending visit to Texas. IBWC was figuratively thrown out of their office. The Act is now in the fourth year of a ten-year authorization. This is but one example of how state and federal claims to water rights can collide.

The federal government has intervened in state water policy to finance and construct large-scale water projects, legislatively secure water earmarked for Indian reservations, finance settlements, and protect public water uses. It also has determined the constitutionality of state restrictions on interstate water transfers and resolved interstate water disputes.

These functions often have been undertaken in response to either a state request or a federal obligation to protect public interests neglected in state water allocations. Many federal agencies deal with water issues along with federal laws including the National Environmental Policy Act, the Clean Water Act, the Endangered Species Act, and federal claims to reserved water rights. These activities will secure the federal governments' continuing role in dealing with water in every state. And states will continue to resist.

## State v. Tribal Water Rights

> "All things are connected, like the blood that runs in your family...
> The water's murmur is the voice of my father's father.
> The rivers are our brothers.
> They quench our thirst.
> They carry our canoes and feed our children.
> You must give to the rivers the kindness you would give to any brother."
> —Suquamish Chief Seattle

Tribes in New Mexico have determined some practical issues for developing tribal regulatory and administrative systems. They include pending legislation, court decisions, administrative actions, tribal water codes and related laws, and leasing settlement water. The Tribes also acknowledge issues related to field investigations and enforcement, administrative hearings and appeals, and the role of the tribal court and tribal council.

The federal government with the help of New Mexico's congressional delegation has made great strides in settling water disputes with Tribes. But because the issues are physically far reaching and long lasting with growing human populations on and off tribal lands, disputes are anticipated in the future.

## Endangered Species

New Mexico has forty-two plants and animals that are federally designated as threatened or endangered. Twenty-two of the twenty-nine animal species are dependent on live or free-standing water for habitat, while two of the thirteen plants are dependent on live or free-standing water for habitat. The twenty-two threatened and endangered animals include thirteen fish, two bird, six crustacean, and one amphibian species. There are petitions and law suits by concerned citizen groups to add more every year. Many species, such as the Socorro isopod, have little chance of ever being delisted unless they go extinct. This animal's worldwide location is one thermal spring near Socorro, New Mexico, where the temperature is between 88 and 90° F. In the late 1970s, the water flowing from the thermal spring was diverted for the development of a spa. The spa has since gone out of business; however, the diversion confined the animal to two small concrete-lined troughs, where it appears to be persisting. This is the only site in the wild where Socorro isopods can be found. A refugium population is being maintained at a facility near Socorro to provide stock for reintroduction into the wild in the event that the wild population is lost due to drought, contaminants, or habitat degradation.

The federal Endangered Species Act is controversial. While some proponents have the best of intentions in saving, preserving, and enhancing threatened and endangered species, others are accused of using this act for job security, perpetuation of environmental organizations, and control of private lands and public land access. Wars over endangered species are expected to continue, possibly forever.

> "Always drink upstream from the herd."
> —Will Rogers

## Gila River Settlement

In 2004, President Bush signed the Arizona Water Settlements Act that provides up to 14,000 acre-feet of water per year for New Mexico and $66 to $128 million in non-reimbursable federal funding. Expenditures from the fund must meet a water supply demand and be approved by the New Mexico Interstate Stream Commission in consultation with the Southwest New Mexico Water Planning Group.

The Interstate Stream Commission outlined the process for determining water use and money expenditure in July 2006:

"The direction from both the New Mexico Interstate Stream Commission and the Governor of New Mexico is to use the best available science and information, coupled with a full and inclusive public involvement process, to both protect the unique and valuable ecology of the Gila Basin and to provide for present and future water needs. By 2014 New Mexico must give notice to the Secretary of the Interior how, or if, New Mexico wishes to utilize its benefits under the Act. Notice to the Secretary must be based on sound science and reasoning. The goal of this planning and decision process must be to provide the citizens of southwestern New Mexico the information and data they need to come to an informed and considered decision and to get them that information in a timely manner."[2]

"The Act requires full compliance with all provisions of federal environmental mandates including the National Environmental Policy Act and the Endangered Species Act. The impacts on state and federally listed species from any use of the funds or development of the water that New Mexico gained in the 2004 Arizona Water Settlements Act is a critical consideration in any decision on how to utilize those benefits."

"There is only a short time to complete the required studies and planning/decision process. The key to a successful planning and decision process is collaborative management by the Gila-San Francisco Coordinating Committee (GSFCC). The GSFCC is composed of representatives of the U. S. Fish and Wildlife Service, the Bureau of Reclamation, the Southwest Water Planning Group (representatives of local governments in southwest New Mexico), the New Mexico Interstate Stream Commission, and the New Mexico Office of the Governor."

"The GSFCC will be responsible for coordinating this initial study of possible impacts to endangered species. The Technical and Public Involvement Subcommittees, independent science forums, public workshops and meetings, and other necessary studies and work will provide input to the GSFCC."

"Currently, the Technical Subcommittee is about to present a prioritized list of studies that they recommend to the GSFCC to be completed in the next two years. Additionally, the first Science Forum was held in October 2005 with six internationally renowned scientists discussing watersheds around the world with similar decisions that we have on the Gila."

"A decision on how, or if, to best utilize the benefits New Mexico received in the 2004 Act will be successful only if it is applied far into the future, not only in the present. More than any other element, the success of this process is contingent on the full and collaborative involvement of federal partners, stakeholders, decision-makers, and the public."

Is it possible by 2014? There wasn't much agreement by 2011. All stakeholders are diverse with many opposing uses. Municipal, farming, and mining interests would like to divert the water for out-of-stream use. Environmental and wildlife groups would like to leave the water in the river for instream uses such as wildlife sustainability. The first group sees this as lost economic opportunities and lost water that eventually ends up in Arizona. After the Interstate Stream Commission makes its decisions, litigation in the courts is likely to follow.

**Domestic Well Problems**

New Mexico's human population continues to grow from the natural birth rate and people moving in from other states and countries even in times of slowed economies. Housing developers find water for new homes from wells, mutual domestic water providers, or municipal systems. The expansion areas often do not have access to the latter two, so new wells must be drilled. State laws allow a home site to use two acre-feet each year. This goes back to early statehood when home sites were large and included vegetable gardens, orchards, and farm animals. People may have a basic human right to enough water for drinking and maybe washing, but that much water is not needed by most home sites today. Two acre-feet per year is equivalent to 1,785 gallons per day!

A developer can buy 100 acres of land with water rights, sell the water rights for a nice profit, and divide the land into 100 one-acre home sites. The new home owners can drill a well for each home site. This gives the new home site owner senior water rights over everyone in the basin like farmers and municipalities except other home site owners. And their rights are equal to other home owners, regardless of how long they have been there. There are thousands of new domestic well permits issued each year, and the state engineer has little choice but to issue them. Because recent New Mexico legislatures have failed to change the law (their tenure is subject to the votes of citizens), the state engineer has attempted to control new domestic well permits through regulation. This is being challenged in New Mexico courts and will probably continue to be challenged well into the future.

**Water Quality**

Septic systems are attached to tens of thousands of individual homes in New Mexico. They are highly efficient for treating wastewater from a home, if they are properly designed, installed, or maintained. However, many are decades old and were not properly designed, installed, or maintained for efficient treatment. In improperly designed, installed, and maintained systems, sewage without enough treatment can move for hundreds of feet from a home to contaminate the environment or threaten public health. Improperly functioning septic systems can be difficult to detect, and their effects can last for years.

New Mexico is the fastest growing dairy state and the number of dairies has doubled in the last ten years. It has approximately 172 dairies, with the largest average herd size (2,088) in the nation. New Mexico is currently ranked 7th in the nation for milk production and 8th in the nation for cheese production. Everyday an average cow produces 6 to 7 gallons of milk and 18 gallons of manure. New Mexico has 300,000 milk cows. That totals 5.4 million gallons of manure produced in the state every day. Dealing with this waste is the dairy industry's greatest environmental challenge. Farms dispose of waste in two ways. First, waste from the milking barn flows into a plastic- or clay-lined lagoon where the liquid evaporates. Second, waste from the feedlots where the cows live is collected and used as fertilizer for grain crops.[3]

The New Mexico Environment Department reports that two-thirds of the state's dairies are contaminating groundwater with excess nitrogen from cattle excrement.[4] Either the lagoons are leaking, or manure is being applied too heavily on farmland. The effects are large and long-term. This will probably be a concern for many years to come.

The Federal Clean Water Act has been controversial since it was heavily amended in 1972. Before that, it mostly implied that it was not nice to pollute. The act applies to all navigable streams and their tributaries. The 1972 amendments pointed to non-point source pollution for the first time. It originally required Best Management Practices be applied to all waters under its jurisdiction. Then a Total Maximum Daily Load (TMDL) limit for pollutants was applied. Confusion immediately set in over issues involving natural levels versus human-caused levels, annual fluctuations, and tolerable levels. Federal and state agencies are still trying to enforce regulations when their TMDL limits cannot be scientifically defended.

The Clean Water Act also applies to the destruction and creation of wetlands. At first, wetlands could not be destroyed, but this was later changed to a no-net-loss arrangement, where every acre that was destroyed to build a mall, for example, required the construction of a new wetland of equal size. For point sources of pollution, such as water coming from a pipe, a National Pollutant Discharge Elimination System (NPDES) permit is required. The most recent controversy involves the extent of applicability. The act refers to navigable waters and its tributaries. Judges in recent years have excluded many water bodies, such as playa lakes that do not feed into any stream. Also of question are rivers, such as the Mimbres in New Mexico, which are not subject to the ebb and flow of the tide or are not being used currently or in the past, or potentially in the future, to transport interstate or foreign commerce. There are several efforts to have the U.S. Congress change the act to include all waters in the United States. Irrigation districts oppose such efforts because they fear their fields and delivery systems will be included in such a broad definition.

Under the Clean Water Act, there are several levels of protection available for rivers and lakes. The highest level of protection comes from a designation as an Outstanding National Resource Water (ONRW). ONRW status means that no activity is permissible if it will result in lower water quality than already exists in the affected water. The New Mexico Water Quality Control Commission gave ONRW protection to more than 700 miles of rivers and streams, more than two dozen lakes, and thousands of acres of wetlands in U.S. Forest Service wilderness areas in the state. The designations also include the Río Santa Barbara and the waters of the Valle Vidal.[5] State officials said that the "no pollution" standard will prohibit any activities, including cattle grazing, which would degrade water quality. The commission's decision was not, however, unopposed. The New Mexico Cattle Growers Association and others sought to derail the designations with litigation subsequently rejected by the New Mexico Supreme Court. Some ranchers and water associations have criticized the designation as being too broad and may continue to fight it through the state legislature.

**Desalination and "New Water"**

Desalination is an emerging technology to provide more drinkable water. This technology currently provides two percent of the world's drinking water. In 2010, there were 15,233 desalination plants in the world, with 244 plants under construction and 700 new plants commissioned.[6] Existing plants provide 15.8 billion gallons per day with 61 percent coming from sea water and 39 percent coming from inland water. About 75 percent of New Mexico's groundwater is brackish and much of it could be desalinated. A major concern is disposal or management of the by-product of desalination, called concentrate. New technologies need to be developed to address this problem. Possibilities exist, but few have been developed to be practical, economical, and environmentally friendly. Possibilities being investigated include surface water discharge, sewer discharge, deep well injection, evaporation ponds, rapid infiltration, oil field injection, and solar ponds. Other methods might be land application and irrigation, zero liquid discharge, aquaculture, wetland creation and restoration, constructed wetlands, and stormwater blending. Concentrate could also be used for recreation purposes, transport of mineral resources, subsurface storage, feedstock for sodium hypochlorite generation, cooling water, dust control and de-icing, and making biodiesel from algae. The best solution for concentrate disposal in each situation is yet to be determined.

**Extreme Climatic Events**

Floods, droughts, and fires have affected New Mexico since humans began migrating to the state many millennia ago. Some people argue that these events have become more frequent and severe in the last 100 years. Others contend that the events are not more frequent and severe but rather they affect more people because of a growing population.

Humans have the most influence over fires that are less frequent but more catastrophic than 100 years ago. We praise fires all winter as being natural functions in the ecosystem, and then fight to extinguish them all summer. This fire prevention results in high fuel buildups that produce extremely hot and long lasting fires when they do occur. Pre-settlement fires were more frequent, relatively less intense, and shorter lasting and resulted in extensive and diverse grasslands. Over longer periods of time, these grasslands prevented flooding and kept water quality high. Until prescribed fires become more frequent and widespread, they will continue to have the potential to be catastrophic with tremendous impacts on our watersheds and downstream users, who will have to deal with higher flows and lower water quality.

Floods have occurred in every stream and arroyo in New Mexico. Floods with the highest recorded flows in New Mexico occurred in 1941. As a result Lake McMillan near Carlsbad was filled with sediment and abandoned for irrigation purposes. Cochiti Dam upstream from Bernalillo was constructed and operated by the U.S. Army Corps of Engineers to control flood flows. Dams in New Mexico have been built for different purposes like storing irrigation water but they have also served well in capturing flood waters. Indeed, the Rio Grande in the Mesilla Valley changed course across the valley every few years before Elephant Butte Dam was built in 1916. Many arroyos in New Mexico have been dammed and maintained by the U.S. Army Corps of Engineers for flood control. Today many need additional maintenance and more need to be built. This was apparent in 2006 when a flood came down an un-dammed arroyo into the Village of Hatch and damaged over 400 homes. Volunteers worked for weeks in relief and restoration efforts. Their efforts were followed by state government assistance and finally the Federal Emergency Management Agency (FEMA). New Mexico is not completely prepared for floods and their effects.

Floods are not nearly as widespread and frequent as droughts. Because New Mexico is mostly arid with some semi-arid lands, annual precipitation over most of the state is below the mean value for the entire state. With storage reservoirs, water from wet periods can be saved to help get through dry periods. As an example, Elephant Butte Reservoir holds about a three-year supply when full. However, prolonged drought results in inadequate storage so that surface supplies for irrigation must be supplemented with groundwater pumping to raise a complete crop. Groundwater pumping can reduce the flow of rivers, impacting deliveries to meet interstate compacts and international treaty deliveries. Priority administration or a priority call refers to the temporary curtailment of junior water rights in times of shortage, so that more senior water rights can be served by the available water supply.[7] As shown on the Pecos River, a priority call to reduce pumping is difficult, unpopular, expensive, and inefficient. New

Mexico's Interstate State Commissioner will continue to face similar problems in the future.

> "Until I came to New Mexico, I never realized how much beauty water adds to a river."
> —Mark Twain

**Other Environmental Concerns**

No one wants a sewage treatment plant near their home. In spite of all the engineering procedures and claims to control odors, they still smell. And land values adjacent to a sewage treatment plant go down or don't rise nearly as fast as they would without the facility. With a growing human population, existing plants will need to be expanded, and more sewage treatment plants will be needed in New Mexico. Controversy and litigation are sure to follow.

Instream flows have been recognized as a beneficial use in New Mexico. They are usually desired by the environmental community to sustain the biology of streams. Currently, most major stream flows in New Mexico are regulated by reservoir releases and change drastically between the irrigation and non-irrigation seasons. This affects the flows' biology. To ensure more instream flow, a water right must be purchased. To provide meaningful protection, water right purchases would require millions to hundreds of millions of dollars for a stream. Thus, environmental entities interested in sustained instream flow often seek water through less expensive voluntary and involuntary forfeitures. If unsuccessful, they shift to purchasing water rights, typically with someone else's money. One provider of such funds has been the federal government. This pattern can be expected to continue in the future.

For a century it has been claimed that clearing large areas of trees and shrubs in forests and riparian areas will increase surface flows and groundwater recharge. However, many of the trees and shrubs being cleared provide valuable services and goods, such as wood products, beauty, wildlife habitat, and recreation. The most-targeted species have been the exotics, salt cedar and Russian olive, which grow along nearly every stream and streamlet below the conifer belt in New Mexico. Thousands of acres have been treated in New Mexico to control these species, with hopes of increasing stream flows. As long as scientific evidence to substantiate increased surface flows or aquifer recharge continues to be lacking, this controversy will continue.

"No river should ever reach the sea."
—Joseph Stalin

"In a mucked up lovely river, I cast my little fly.
I look at that river and smell it and it makes me wanna cry.
Oh to clean our dirty planet, now there's a noble wish,
and I'm puttin my shoulder to the wheel
'cause I wanna catch some fish."
—Greg Brown, "Spring Wind" from *Dream Café*, 1992

## Adjudication

An adjudication is a lawsuit, usually filed by the state, for determining who owns what water rights in a river system. Each different stream system has its own separate lawsuit. When the Office of the State Engineer (OSE) files an adjudication suit, each water rights owner becomes a defendant and must establish the amount and extent of his or her water rights. This includes well owners, towns, municipalities, tribes, acequias, pueblos, irrigation districts and their members, and the United States (if the adjudication involves federal water rights on federal lands).

Adjudications often take decades to complete. In any adjudication, there can be hundreds or even thousands of defendants, and each water rights owner has separate elements of his or her water rights that must be defined. In addition, the judges or "special masters" are often involved in multiple adjudications, and so there can be long periods in which the court's attention is focused on a different adjudication. The OSE, as the "plaintiff," can strongly influence the amount of activity or inactivity in each adjudication. Also, if an adjudication deals with new or complicated legal issues, it can take years to conduct hearings and to gather briefs from the different parties in order to come to a resolution. It has been estimated that the adjudication process will take several hundred more years to complete in New Mexico at the present rate of progress. Because of its nature, disputes and wars will persist.

## Service Area Disputes

New Mexico has hundreds of small mutual domestic water providers. Many of these are adjacent to municipalities. As these municipalities grow and annex adjacent lands, they often annex areas that are served by these mutual domestic water providers. Sometimes the mutual domestic water providers merge with the municipality's water system and sometimes they continue to be independent. Municipalities are often better able to find financial resources to upgrade infrastructure and comply with changing environmental regulations. Although small water providers suffer from low profitability, they still cherish autonomy and local control. Disputes over service areas will continue as long as municipalities continue to expand by annexation.

## Transfer of Water from Agriculture Lands to Municipal, Industrial, and Environmental Uses

Transferring water from agriculture lands to municipal, industrial, and environmental uses can be compared to regions of a chile pepper where heat increases from the tip towards the stem. Transferring water from agriculture within a local water district or within a basin between districts is usually mild to warm to hot. Transferring water from agriculture to environmental uses is always hot. Interbasin and interstate transfer of water ranges from boiling to searing to explosive. Water marketing is sometimes referred to as horse trading. This is the nature of cultures and politics. They are not expected to change in the future.

## Many Entities Still Looking for Cheap or Free Water

Water is cheap. Most households pay a third to half as much for a month's water supply to the home as they do for cable television, and even less compared to cell phone charges. People argue that water for drinking and washing is a basic human right and should be free. Yet they will pay $3.95 for a liter of water in their room at a major hotel. This is equivalent to $4,870,823 per acre-foot. Most people don't realize how much water they use in their homes every day (about 150 gallons per person). Their accumulated monthly water bill is low compared to what it probably will be in the future. Raising water rates by utilities is always controversial and unpopular. Rates will increase in the future. Water is the most critical resource issue of our lifetime and our children's lifetime.

> "The health of our waters is the principal measure of how we live on the land."
> —Luna Leopold

**Groundwater Mining and Interbasin Transfers**

If surface water supplies decline in the next decades or century as some scientists have predicted, then the use of other sources, such as groundwater, could significantly increase. Large aquifers in New Mexico are not currently being used much but are being considered. These include the aquifers under the St. Augustine Plain, the Estancia Basin, the Tularosa Basin, the Salt Basin, and Otero Mesa, among others. Most of the water in these aquifers is old, in that it was placed there during the Pleistocene era (beginning about 2.8 million years ago) or even before. Pumping would result in this water being "mined," generally meaning removed and not replaced. This process is unpopular among some, because it is not sustainable. However, the Tularosa Basin is estimated to have more than one billion acre-feet of water that is currently being little used because it is brackish. Much of it contains salt concentrations between 1,000 and 2,000 parts per million, which makes desalination quite feasible in cases where transport to areas of need is economically available. Those one billion acre-feet are enough to supply New Mexico's present municipal and household needs for 5,000 years. Interbasin transfers raise concerns that probably will continue into the future.

**Conclusions**

As Judge Valentine said in Chapter 2, "Future disputes may be as contentious as past or present ones, and it is likely that the judiciary will be involved in resolving water disputes." Although poorly documented, Mark Twain is credited with coining the phrase, "Whiskey is for drinking, water is for fighting." Will New Mexico's water woes ever be solved and its fights ever stop? As the old saying goes "The wheels of change turn slowly," and this applies to New Mexico.

"The crisis of our diminishing water resources is just as severe as any wartime crisis we have ever faced. Our survival is just as much at stake as it was at the time of Pearl Harbor, or the Argonne, or Gettysburg, or Saratoga."
—Jim Wright, U.S. Representative, *The Coming Water Famine*, 1966

## Notes

1. New Mexico Acequia Association website www.lasacequias.org
2. New Mexico Interstate Stream Commission Briefing on Upper Gila River Settlement Decision Process www.ose.state.nm.us/PDF/ISC/BasinsPrograms/GilaSanFrancisco/BriefingPacket-7-3-2006.pdf
3. Farm-To-Consumer Legal Defense Fund www.ftcldf.org/news/news-11Dec2009-6.html
4. National Public Radio www.npr.org/templates/story/story.php?storyId=121173780
5. New Mexico Environment Department's Surface Water Quality Bureau www.nmenv.state.nm.us/swqb/ONRW
6. Desalination and Water Purification Technology Road Map www.sandia.gov/water/desal/docs/DesalRdmap04a.pdf
7. Office of the State Engineer www.ose.state.nm.us/faq_index.html#5

# Epilogue

## Michael L. Connor

*I* am honored to be asked to write an epilogue for this book—*One Hundred Years of Water Wars in New Mexico 1912–2012*. As a New Mexican (including a New Mexico State University graduate), I feel very fortunate to have had the opportunity in my career to participate in, and help bring some level of resolution to several of the disputes written about in this book. Foremost among my opportunities were the eight years I worked for U.S. Senator Jeff Bingaman on the Senate Energy and Natural Resources Committee. As both Chairman and Ranking Member of the Committee, Senator Bingaman always understood the importance of water to New Mexico, and focused a large amount of his time and energy on addressing the state's present and future water challenges.

As Commissioner of the Bureau of Reclamation, I now work for Secretary of the Interior Ken Salazar, a Coloradan (with strong family connections to New Mexico), who has a wealth of experience in western water issues and believes very strongly that Reclamation can and should play a significant role in helping to address the West's water challenges over the next 100 years. Of course, in this position I continue to be immersed in New Mexico's water issues, given Reclamation's role in helping New Mexico's development during this past century. Reclamation has an ongoing responsibility to operate and maintain projects in each of New Mexico's major river basins, and was also recently charged with new responsibilities to build significant new infrastructure that will implement several recent Indian water rights settlements. This infrastructure will provide significant benefits to many New Mexico tribes and adjacent non-tribal communities, for years to come.

Reclamation's current goals are simple but important—promoting certainty and sustainability in the use of the West's limited water resources. Notwithstanding the non-controversial nature of these goals and the overall benefits provided by these projects, we are still dealing with water in the West—a subject that is controversial by its very nature. As noted in several of the chapters of this book, Reclamation has been involved in many of New Mexico's water wars, and that trend looks as though it will continue. About a month before I was asked to write this epilogue, on August 8, 2011, New Mexico's Attorney General filed suit against Reclamation regarding the accounting of water in Rio Grande Project Storage under the Rio Grande Compact and the 2008 Rio Grande Operating Agreement referenced in Chapter 16. As Em Hall accurately predicted, the future water wars will be a continuation of all the past wars (see Chapter 16). My reaction to New Mexico's suit is similar to that of Las Cruces officials when El Paso sued in 1980. Reclamation is certainly willing to "work something out" and overall, I think the suit is "unfortunate" (see Chapter 14). Instead of continuing down a constructive path of dialogue regarding Reclamation's operation of the Rio Grande Project, the lawsuit threatens to squelch discussions in favor of depositions, interrogatories, and legal motions that prolong uncertainty. Hopefully, this lawsuit is just a temporary distraction, and we can avoid it becoming the next lengthy chapter in the second 100 years of New Mexico water wars.

In the last chapter of this book, Karl Wood discusses the prospects for future water wars. He highlights simmering tensions that are likely to boil over, as well as outlining new pressures on New Mexico's limited water resources that are likely to raise new tensions. Karl's assessment is spot-on as usual. Nonetheless, if many of his predictions turn out accurate, the certainty and sustainability that Reclamation and many others are seeking will become increasingly elusive. The stories contained in this book have made plainly clear that legal proceedings typically take a long time to play out and even when decisions get rendered, they address only a fraction of the issues that need resolution in order to effectively manage the resource for the long-term. Don't get me wrong—I enjoy a good and interesting story. New Mexico's water wars and courtroom battles provide an abundance of those. However, given the scarce nature of New Mexico's most precious resource, increasing demands on its use, and the very real prospects that climate change is limiting its availability, a shared vision of the common strategies needed for good water management will be at a premium. When conflicts arise, good faith negotiation and compromises will be even more critical in the future if the certainty needed to sustain our economy and the environment is to be realized. I like to think that as our water management challenges get tougher in the future, New Mexicans will pull together and make decisions that will allow us to continue to thrive and prosper in this great state.

As a final note, I'd just like to acknowledge that my tenure at the helm of the Bureau of Reclamation has given me a strong appreciation for my predecessors at the federal level, as well as the lead water resource managers at the state level. I never had the pleasure of meeting Steve Reynolds, but I certainly hold his accomplishments and longevity in high regard. John D'Antonio recently completed a successful run as New Mexico State Engineer, and should be proud of the water-related accomplishments secured by the state during his tenure. Of course, Eluid Martinez, a co-author of one of the chapters in the book, accomplished much as both a New Mexico State Engineer and Commissioner of the Bureau of Reclamation. Even as I mention and highlight the service of these individuals, I would note the obvious—that New Mexico has been blessed with a large number of talented water professionals and yes, water lawyers, who have kept things interesting over the past 100 years and solved a number of water resource problems along the way. My goal in this position is to simply try, to the extent possible, to minimize the interesting part of the stories, and work with all parties on the solutions that will hopefully define the 21$^{st}$ Century.

# Contributors

**James C. Brockmann** practices water law for the firm of Stein and Brockmann, P.A. He works primarily in New Mexico, but has worked with clients in Kansas, Wyoming, Nebraska, and Texas. He undertook two successful trips to the U.S. Supreme Court on behalf of the State of Nebraska.

**Calvin Chavez** retired from the Office of the State Engineer after twenty-six years of service, the last ten years serving as the District 4 manager in Las Cruces. He died shortly after retiring and had just previously submitted a draft chapter for this book.

**Michael L. Connor** was confirmed Commissioner of the Bureau of Reclamation by the United States Senate on May 21, 2009. Connor has more than 15 years of experience in the public sector, including having served as Counsel to the U.S. Senate Energy and Natural Resources Committee since May 2001.

**Charles T. DuMars** is professor emeritus at the University of New Mexico School of Law. He is currently with the law firm Law & Resource Planning Associates, P.C. in Albuquerque, where he practices water and environmental law.

**Em Hall** recently retired from the University of New Mexico School of Law. He is the author of *Four Leagues of Pecos: A Legal History of the Pecos Grant from 1800 to 1936* (1984) and *High and Dry: The Texas-New Mexico Struggle for the Pecos River* (2002). Prior to joining the UNM law faculty in 1983, he spent seven years at the Office of the State Engineer.

**Linda G. Harris** covered the El Paso hearing as a writer for the New Mexico Water Resources Research Institute. She is the author of four books on New Mexico and lives in Las Cruces. Her chapter is based on a 1990 publication of the same name, which she wrote with contributing authors Robert C. Czerniak, Richard A. Earl, and William J. Gribb, all of New Mexico State University's Geography Department. The facts and figures cited here date from the time of the hearing.

**John W. Hernandez** is professor emeritus, Department of Civil Engineering, New Mexico State University. He has extensive experience in water resources management issues, particularly in water quality and at various times throughout his career worked for the Office of the State Engineer.

**Eluid L. Martinez** is a distinguished engineer with extensive experience in water resource planning. He served in the New Mexico Office for the State Engineer for 23 years, including holding the post of State Engineer. Nominated by President Clinton and confirmed unanimously by the U.S. Senate, he served as Commissioner for the Bureau of Reclamation from 1995 to 2001.

**Kay Matthews** is editor of *La Jicarita News*, a community advocacy newspaper based in northern New Mexico. She lives in El Valle on a 10-acre farm where she grows vegetables, fruit, and hay, and serves as a commissioner on the Acequia Abajo de El Valle. She will continue to cover the Lower Rio Grande adjudication (www.lajicaritanews.org). Another resource that provides extensive background on the Rio Grande Project is Ira G. Clark's *Water in New Mexico, a History of its Management and Use*.

**John Nichols** is author of the New Mexico Trilogy and many other works including *The Milagro Beanfield War*, which was made into a movie by the same name. He has lived in northern New Mexico since 1969.

**Sylvia Rodríguez** is professor emerita of anthropology at the University of New Mexico and author of *Acequia: Water Sharing, Sanctity, and Place*. She is engaged in collaborative and participatory research on acequias.

**Richard Simms** was general counsel to Steve Reynolds from 1976 to 1982. He has been extremely active and successful in representing private interests and states in the West in litigation over Indian and federal reserved water rights and the scope of state court jurisdiction in McCarran Amendment litigation. He has also represented New Mexico successfully in three original actions between states in the United States Supreme Court, as well as two other states in original actions.

**Jay F. Stein** is a shareholder with Stein & Brockmann, P.A., practicing water law throughout the State of New Mexico and regionally. He has worked extensively on interstate and intrastate water issues. He served as a Special Assistant Attorney General with the New Mexico Office of the State Engineer and Interstate Stream Commission before entering private practice.

**John W. Utton** is a water lawyer living in Santa Fe. He has represented the State of New Mexico in settlement of the Navajo Nation's claims to the San Juan River. He has also represented acequia associations in tribal water rights negotiations and settlements on the Rio Chama and Rio Jemez. In the *Aamodt* case, he represents Santa Fe County.

**Jerald A. Valentine** was appointed judge in the Lower Rio Grande Adjudication in late 1995 and remained in that capacity until he retired in late 2010. He was appointed District Judge, Division IV, 3rd Judicial District Court in 1993 and Chief Judge, 3rd Judicial District Court from 1999–2002, and again from October 2008 to 2010.

**M. Karl Wood** retired as director of the New Mexico Water Resources Research Institute in 2010 after a ten-year tenure. Prior to becoming director, he was a professor of animal and range sciences at New Mexico State University.

www.ingramcontent.com/pod-product-compliance
Lightning Source LLC
Chambersburg PA
CBHW020833160426
43192CB00007B/624